AF790593

Lattice-mismatched Epitaxy for Fabricating HgCdTe Infrared Materials and Detectors

Online at: https://doi.org/10.1088/978-0-7503-3443-3

IOP Series in Sensors and Sensor Systems

The IOP Series in Sensors and Sensor Systems includes books on all aspects of the science and technology of sensors and sensor systems. Spanning fundamentals, fabrication, applications, and processing, the series aims to provide a library for instrument and measurement scientists, engineers, and technologists in universities and industry.

The series seeks (but is not restricted to) publications in the following topics:

- Advanced materials for sensing
- Biosensors
- Chemical sensors
- Industrial applications
- Internet of Things (IoT)
- Lab-on-a-chip
- Localization and object tracking
- Manufacturing and packaging
- Mechanisms, modeling, and simulations
- Microelectromechanical systems/nanoelectromechanical systems
- Micro and nanosensors
- Non-destructive testing
- Optoelectronic and photonic sensors
- Optomechanical sensors
- Physical sensors
- Remote sensors
- Sensing for health, safety, and security
- Sensing principles
- Sensing systems
- Sensor arrays
- Sensor devices
- Sensor networks
- Sensor technology and applications
- Signal processing and data analysis
- Smart sensors and monitoring
- Telemetry

Authors are encouraged to take advantage of electronic publication through the use of color, animations, video, and interactive elements to enhance the reader experience.

A full list of titles published in this series can be found here: https://iopscience.iop.org/bookListInfo/iop-series-in-sensors-and-sensor-systems.

Lattice-mismatched Epitaxy for Fabricating HgCdTe Infrared Materials and Detectors

Wen Lei
Department of Electrical, Electronic and Computer Engineering, The University of Western Australia, Perth, Australia

IOP Publishing, Bristol, UK

© IOP Publishing Ltd 2025. All rights, including for text and data mining (TDM), artificial intelligence (AI) training, and similar technologies, are reserved.

This book is available under the terms of the IOP-Standard Books License

No part of this publication may be reproduced, stored in a retrieval system, subjected to any form of TDM or used for the training of any AI systems or similar technologies, or transmitted in any form or by any means, electronic, mechanical, photocopying, recording or otherwise, without the prior permission of the publisher, or as expressly permitted by law or under terms agreed with the appropriate rights organization. Certain types of copying may be permitted in accordance with the terms of licences issued by the Copyright Licensing Agency, the Copyright Clearance Centre and other reproduction rights organizations.

Permission to make use of IOP Publishing content other than as set out above may be sought at permissions@ioppublishing.org.

Wen Lei has asserted his right to be identified as the author of this work in accordance with sections 77 and 78 of the Copyright, Designs and Patents Act 1988.

ISBN 978-0-7503-3443-3 (ebook)
ISBN 978-0-7503-3441-9 (print)
ISBN 978-0-7503-3444-0 (myPrint)
ISBN 978-0-7503-3442-6 (mobi)

DOI 10.1088/978-0-7503-3443-3

Version: 20251201

IOP ebooks

British Library Cataloguing-in-Publication Data: A catalogue record for this book is available from the British Library.

Published by IOP Publishing, wholly owned by The Institute of Physics, London

IOP Publishing, No.2 The Distillery, Glassfields, Avon Street, Bristol, BS2 0GR, UK

US Office: IOP Publishing, Inc., 190 North Independence Mall West, Suite 601, Philadelphia, PA 19106, USA

Contents

Preface

Semiconductor heterostructures are the building blocks for modern electronic and optoelectronic devices, which are formed by epitaxial growth of one material on another. However, each material, by nature, has its own unique physical properties, in particular crystal lattice constant, which is usually not well matched across the heterostructure interface. Typically, this lattice mismatch between materials results in the generation of misfit dislocations in the vicinity of the interface which form threading dislocations while propagating to the surface. These dislocations deteriorate the physical properties and thus the device yield and performance. As a result, current high-performance heterostructure devices are reliant on lattice-matched material systems in order to avoid the generation of misfit and threading dislocations. However, the number of lattice-matched heterostructured material systems is very limited, and cannot meet the increasing performance demand of future industry applications. Therefore, significant attention has been devoted to heteroepitaxy with large lattice mismatch, such as III–V semiconductors (GaAs, InP, GaSb, InSb) on Si, as well as II–VI semiconductors on Si, Ge, and III–V substrates. The principal challenge for these heteroepitaxial growth processes relates to the large lattice mismatch and thus higher density of threading dislocations generated. Although significant progress has been made in high-quality heteroepitaxy, the performance of electronic devices based on lattice-mismatched heteroepitaxy is still much lower than that of counterparts grown on lattice-matched substrates due to the high dislocation density generated during heteroepitaxy. Further effort is needed to achieve high-quality lattice-mismatched heteroepitaxial growth with material quality and subsequent device performance comparable to the lattice-matched ones.

This book will take the strategically important II–VI semiconductor HgCdTe as an example, and review and discuss the lattice-mismatched heteroepitaxy for fabricating high-performance HgCdTe materials and infrared detectors. Chapter 1 presents a brief introduction on HgCdTe infrared detectors and their challenges of higher cost and smaller array format, together with the solution—lattice-mismatched heteroepitaxy for fabricating HgCdTe materials and detectors. Chapter 2 presents a theoretical discussion on the growth mechanisms related to lattice-mismatched heteroepitaxy and general approaches to reduce the dislocations within. Chapter 3 is dedicated to the progress of heteroepitaxial growth of high quality CdTe and HgCdTe layers on various lattice-mismatched alternative substrates including Si, Ge, GaAs, and GaSb substrates. Chapter 4 focuses on the progress of Van der Waals epitaxy of CdTe and HgCdTe layers on various two-dimensional substrates including graphene, Mica, and other transition metal dichalcogenide (TMD) monolayer substrates. Chapter 5 is dedicated to the progress of HgCdTe infrared detectors grown on various alternative substrates including short-wave infrared (SWIR), mid-wave infrared (MWIR) and long-wave infrared (LWIR) ones. Chapter 6 focuses on an alternative infrared material system—HgCdSe materials on GaSb substrates, for making lower cost, larger format, high-performance infrared detectors. Chapter 7 concludes the whole book by summarizing all the chapters and presenting an outlook and perspective for developing high-performance HgCdTe infrared detectors based on lattice-mismatched heteroepitaxy. We anticipate more innovative and effective heteroepitaxial growth methods and dislocation reduction techniques to come in the future so that high-performance HgCdTe infrared detectors including LWIR ones can be eventually fabricated on alternative substrates with lower cost and larger array format sizes.

This book will introduce the theoretical knowledge and experimental techniques to suppress the generation of dislocations, control the evolution and propagation of dislocations, and reduce the dislocation density in the epitaxial layers. The epitaxial growth of HgCdTe detectors on Si, Ge, GaAs, and GaSb just provides a test vehicle, and all the knowledge and techniques discussed can be applied to other semiconductor material systems and devices. Researchers working in the general area of semiconductor epitaxy and related device applications will benefit from this book by understanding the knowledge and techniques for undertaking high quality lattice-mismatched heteroepitaxy with the goal to achieve high-quality semiconductors on lattice-mismatched substrates, and meet the increasing performance and cost demand of future industry applications.

Wen Lei
July 2025

Acknowledgments

I gratefully acknowledge all those who provided help, support and encouragement during the writing of this book. I would like to begin by expressing my appreciation to the University of Western Australia for providing the facilities and environment for me to undertake my research and work on this book. I would also like to thank the Australian Research Council and other funding agencies for supporting my research on HgCdTe and HgCdSe infrared materials and detectors which form the foundation of this book. I have benefited from the support and collaboration of many colleagues who are actively working on infrared detector technologies, especially Lorenzo Faraone, John Dell, Jarek Antoszewski, Gilberto A. Umana Membreno, Yongling Ren and Charles Musca at the University of Western Australia. Special thanks are to our former and current research fellows and doctoral students who have contributed their time and effort to this book, including Dr. Wenwu Pan, Dr. Imtiaz Madni, Mr. Zekai Zhang, Mr. Shuo Ma and Ms. Fahia Munna. I also would like to thank the relevant publishers and individuals for providing the permissions to reproduce their figures and descriptions for this book.

I would like to extend my special thanks to our publisher – IOP Publishing. IOP Publishing and its staff have been of great help during the process of organising, writing and publishing this book. Their interest in our research work and their patience were of great encouragement. I would like to specifically mention Phoebe Hooper, John Navas, Sarah Armstrong, Ursula Fairhurst, Ashley Gasque, and Debbie Peach for the numerous discussions during the publication of this book. Special thanks are also to the IOP staff who I did not have direct communications with, but provided their effective help and support for publishing this book.

Finally, I would like to thank my family for the encouragement, understanding, and support especially over the past two years. Endless gratitude to you, as I couldn't have done it without you.

Wen Lei
Perth, Western Australia, Australia
November, 2025

Author biography

Wen Lei

Dr Wen Lei is a professor, an ARC (Australian Research Council) Future Fellow, the Program Chair of Electrical and Electronic Engineering, and the Chair of the IEEE Western Australia Joint ESP Chapter (Electron Devices Society, Solid State Circuits Society, and Photonics Society), and leads the Electronic Materials and Devices Research at the Department of Electrical, Electronic and Computer Engineering, The University of Western Australia. His current research mainly focuses on semiconductor materials, devices, and their system applications especially infrared sensors. He was awarded his prestigious ARC Future Fellowship in 2013. He has made a number of 'first in the research field' breakthroughs such as strained superlattices for filtering defects in HgCdTe/CdTe materials, first MBE-grown HgCdSe infrared sensor, and more. He, together with his colleagues, has demonstrated the first Australian-made prototype mid-wave HgCdTe infrared focal plane arrays with commercial format size. He holds a number of patents and has published four book chapters (Springer Publishing), three invited topical review articles, and over 160 high-profile papers in top journals like *Applied Physics Reviews*, *Physical Review Letters*, *Advanced Materials*, *Advanced Functional Materials*, *Nano Letters*, *Advanced Science*, *Small*, *Laser & Photonic Reviews*, *IEEE Transactions on VLSI Systems*, and *Applied Physics Letters*, and more. His research has attracted strong interest from industry partners, leading to a number of ARC Linkage awards. He also received several 'Best Paper' Awards from IEEE journals and IEEE conferences, and is invited regularly to deliver plenary/keynote presentations at premium international conferences. He is also an associate editor/editorial board member for five international journals, a member of Scientific Advisory Board/Technical Program Committee/Organizing Committee for several international conferences and a regular reviewer for various international prestigious journals and funding agencies.

Description

Nowadays, with the development of modern electronic devices, the traditional epitaxial growth of semiconductor materials on lattice-matched substrates is not enough for meeting the increased demand of lower cost and high device performance. In the last two decades, significant attention has been devoted to growing semiconductor materials on lattice-mismatched substrates with lower costs, larger wafer sizes and higher crystal quality, such as GaAs on Si, InP on Si, HgCdTe on Si, and so on. The aim is to have devices with lower cost, but without sacrificing device performance. Therefore, the lattice-mismatched epitaxy growths and their applications in modern electronic devices are one of the most important subjects in the current semiconductor industry, and have attracted significant attention in the past two decades. However, this lattice-mismatched epitaxial growth and its device applications are seriously limited by the high defect density in the materials caused by the lattice mismatch, which needs to be well addressed before achieving the wide application of lattice-mismatched epitaxial growth.

This book will introduce the theoretical knowledge and experimental techniques to suppress the defect generation, control the defect evolution and propagation, and reduce the defect density in the epitaxial layers. The epitaxial growth of HgCdTe detectors on Si, Ge, GaAs, and GaSb discussed in this book just provides a test vehicle, and all the knowledge and techniques can be applied to other semiconductor material systems and devices. The readers will benefit from this book by understanding the knowledge and techniques for achieving high-quality semiconductors on lattice-mismatched substrates.

Chapter 1

Introduction to HgCdTe infrared materials and detectors

As the introduction chapter of this book, we will start with the introduction of infrared spectral bands and their applications, and then discuss infrared detectors—the core electronic components for infrared sensing and imaging applications as well as their performance figures of merit. After that, we will introduce the HgCdTe infrared detectors—a detector technology dominating the high-performance infrared sensing and imaging applications as well as their current status and future challenges. Then we will discuss how to address the challenge of higher cost and smaller array format size, which leads to this book—heteroepitaxial growth of HgCdTe materials on lattice-mismatched alternative substrates. Finally, this chapter will finish by introducing the structure of this book.

1.1 Infrared electromagnetic bands and their applications

Electromagnetic wave/radiation is a self-propagating wave of the electromagnetic field that carries momentum and radiant energy through space. An electromagnetic wave is characterized by a frequency and a wavelength. These two quantities are related to the speed of light by the equation: speed of light (c) = frequency (f) × wavelength (λ). A range of wavelengths of electromagnetic radiation constitute the electromagnetic spectrum. Figure 1.1 presents a typical electromagnetic spectrum in our real world [1]. As shown in figure 1.1, the electromagnetic spectrum can be divided into several spectral bands including radio waves, microwaves, infrared, visible light, ultraviolet, x-rays, and gamma rays. The difference between these spectral bands is their wavelengths/frequencies. In our daily life, people are more familiar with visible light, the wavelength of which ranges from 380 to 700 nm. In this visible spectral band, people can observe different colors such as blue, green,

doi:10.1088/978-0-7503-3443-3ch1 © IOP Publishing Ltd 2025. All rights, including for text and data mining (TDM), artificial intelligence (AI) training, and similar technologies, are reserved.

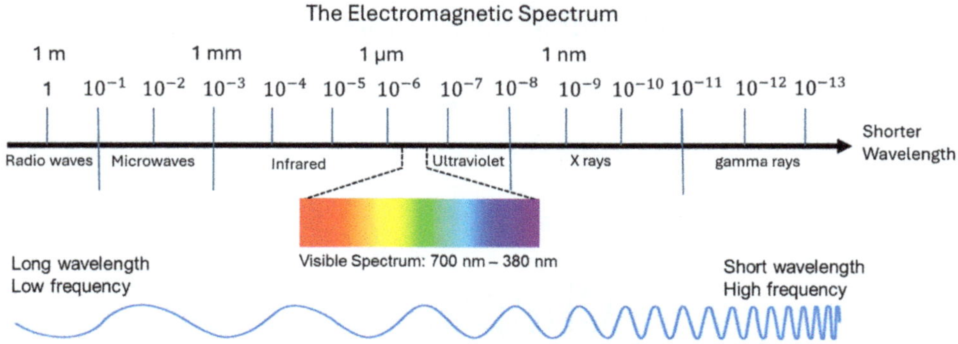

Figure 1.1. Typical electromagnetic spectrum in our real world.

Table 1.1. Wavelength range for infrared spectral bands.

Wavelength band	Abbreviation	Wavelength range (μm)
Near-infrared	NIR	0.7–1
Short-wave infrared	SWIR	1–3
Mid-wave infrared	MWIR	3–5
Long-wave infrared	LWIR	8–12
Very-long wave infrared	VLWIR	> 12

and red. Usually, shorter wavelengths are referred to as 'bluer,' while longer wavelengths are referred to as 'redder.' On the left side of this visible spectral band, there is another strategically important spectral band—the infrared band, the wavelength of which ranges from 700 nm to 1 mm (corresponding to a frequency range from 300 GHz to 400 THz). This infrared spectral band includes most of the thermal radiation emitted by objects near room temperature, and thus are applied widely in our daily life, which will be discussed later in this chapter.

Infrared (IR) wave/radiation was discovered by Sir William Herschel in 1800 when he experimented with a prism and found the colors of light emitted different temperatures, with the temperature increasing beyond the visible red band of the spectrum (>700 nm). This infrared spectral band can be further divided into five sub-bands according to their wavelength ranges: near-infrared (NIR), short-wave infrared (SWIR), mid-wave infrared (MWIR), long-wave infrared (LWIR), and very-long-wave infrared (VLWIR) bands. Table 1.1 lists the wavelength range for each sub-band [2]. Note all objects at temperatures above absolute zero can emit IR radiation. Because energy (E) has the following relationship with electromagnetic wavelength (λ): $E = hc/\lambda$, where h is Planck's constant and c is the speed of light, it means the shorter the wavelength, the higher the corresponding energy and thus the higher the temperature to emit the radiation. As a result, different sub-bands can be used to sense IR radiation emitted from objects with different temperatures. Some examples are as follows: the LWIR band corresponds to IR radiation emitted from objects at room temperature, which can be used for night vision and medical applications; the MWIR band corresponds to IR radiation emitted from objects at several hundred degrees Celsius, which can be used for various defence applications involving imaging or tracking objects including hot engines or plumes; NIR and SWIR bands correspond to IR radiation emitted from objects at much higher temperatures such as sun radiation, which can be used for various defence applications utilizing IR reflection of objects. Therefore, these IR spectral bands have various important applications in various industry sectors. The following showcases two main example applications which are also the main target applications of HgCdTe IR detectors in this book.

Infrared sensing and imaging: Figure 1.2 shows a typical transmission spectrum of an atmosphere [3]. It is observed that those IR sub-bands listed in table 1.1 present high atmospheric transmission. Therefore, these IR bands can be used as atmospheric windows with minimum absorption for terrestrial remote sensing and imaging applications. These IR sensing and imaging applications currently are the most important and useful applications for IR bands, and benefit national security, defence technology, meteorology, and astronomy. This also forms the largest and most important market for IR detectors. Typical example applications of infrared sensing and imaging include: surveillance, remote sensing, astronomy study, defence and security, medical imaging, electrical inspection, and many others [4, 5]. All these indicate the importance of these IR spectral bands.

Infrared spectroscopy study: Apart from being emitted, IR radiation (energy) can be absorbed by molecular vibrations in molecules. More interesting is that different molecular vibrations absorb IR radiation of different energies (frequencies/wavelengths). Thus, the infrared absorption spectrum of a compound is very specific and unique for that compound. In this regard, the IR absorption spectrum is

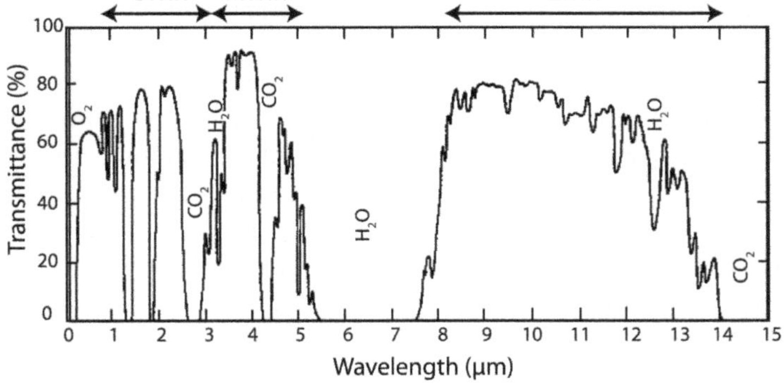

Figure 1.2. Typical transmission spectrum of an atmosphere for a 2000 m horizontal path at sea level. [3] John Wiley & Sons. Copyright © 1969 by John Wiley & Sons, Inc.

sometimes called the 'fingerprint' spectrum of a compound. For example, a number of gases and liquids have a characteristic IR absorption spectrum like volatile organic compounds, which enables identification and detection of these substances with IR detectors, and benefits environmental monitoring, molecular spectroscopy, medical diagnosis, and other applications.

1.2 Infrared detectors and performance choice

For all IR spectral band applications, the most critical electronic components are high-performance IR sensors/detectors and their focal plane arrays which can efficiently detect IR radiation emitted by objects. Generally, IR detectors can be classified into two categories: thermal detectors and photonic detectors (photodetectors), which will be briefly introduced in the following [6, 7]:

Thermal detectors: Thermal detectors operate by sensing changes in the detector's temperature caused by the absorption of IR radiation. This temperature change then affects a temperature-dependent property of the detector, which is measured and converted into an electrical signal. Bolometers and pyroelectric detectors are the two most common types of thermal detectors. Typically, a bolometer works like a resistance thermometer and senses a change in the electrical resistance of a thin film resistor [6–8], while a pyroelectric detector relies on materials' pyroelectric effect and senses temperature-dependent changes in the electrical polarization of the absorbing material [6, 7, 9]. Thermal detectors generally operate at room temperature and are of lower cost, but show relatively lower response speed and detectivity in comparison to photonic detectors. Another feature is that thermal detectors have a blackbody-type wavelength dependence, and thus are sensitive to a broad spectral range of IR radiation, which is different from the photonic detectors.

Photonic detectors: Photonic detectors, also known as photodetectors, are devices that convert light (photons) into an electrical signal. Photonic detectors rely on photon-electron interactions. The incident IR radiation (photons) generate a change in the electrical characteristics (voltage or current) of the material, which is then measured by an electric circuit. Compared with thermal detectors, photonic detectors generally have a higher response speed and detectivity, but are of higher cost. Note that photonic detectors rely on the absorption of IR photons and thus excitation of electrons from the valence band to the conduction band of the absorber materials to form photocurrent/photovoltage. Therefore, they present a wavelength-dependent response (e.g., cut-off wavelength). In addition, many photonic IR detectors require cooling to suppress dark current and thus optimize detector performance. In this regard, IR imaging and sensing systems based on photonic detectors are usually of higher cost than those based on thermal detectors due to the cooling requirements and the use of more complex and/or advanced detector materials and device structures. Apart from the increased cost, cooling requirements also lead to a large size

> and weight of IR systems based on photonic detectors. Despite these limitations, photonic detectors are generally preferred over thermal detectors for applications where performance is an important consideration such as defence and aerospace applications.

Generally, there are three broad categories of IR photonic detectors: intrinsic, extrinsic, and quantum well/dot detectors. Intrinsic detectors rely on the absorption of photons to promote electrons from the valence band into the conduction band of the absorber materials, thereby generating electron-hole pairs to be collected to form photocurrent/photovoltage [6, 7]. Extrinsic detectors rely on the absorption of photons by impurities to promote electrons into the conduction band or to create holes in the valence band by promoting electrons from the valence band into an impurity level [6, 7]. Quantum well and dot detectors rely on the absorption of photons to promote electrons from one quantized energy level into another quantized energy level by leveraging quantum mechanical effect in nanoscale structures [6, 7]. Among these three categories of photonic IR detectors, typically intrinsic detectors present better device performance in similar spectral ranges due to their large light absorption, and they are also of lower cost (relatively easy to fabricate and operate). For the topic of this book—HgCdTe IR detectors—they are typical intrinsic photodetectors. Next, we will take the HgCdTe IR detector as an example for further discussion on intrinsic photodetectors.

For intrinsic HgCdTe IR detectors, there are two main types of detector structures: photoconductor and photovoltaic detector (photodiode). HgCdTe-based photoconductors were the standard detector structure used in the early days (1960s) of the IR industry. A photoconductor behaves like a photosensitive resistor. Figure 1.3(a) shows the schematic working principle of an HgCdTe photoconductor [9]. The IR radiation is absorbed by the HgCdTe material, leading to the generation of electron-hole pairs,

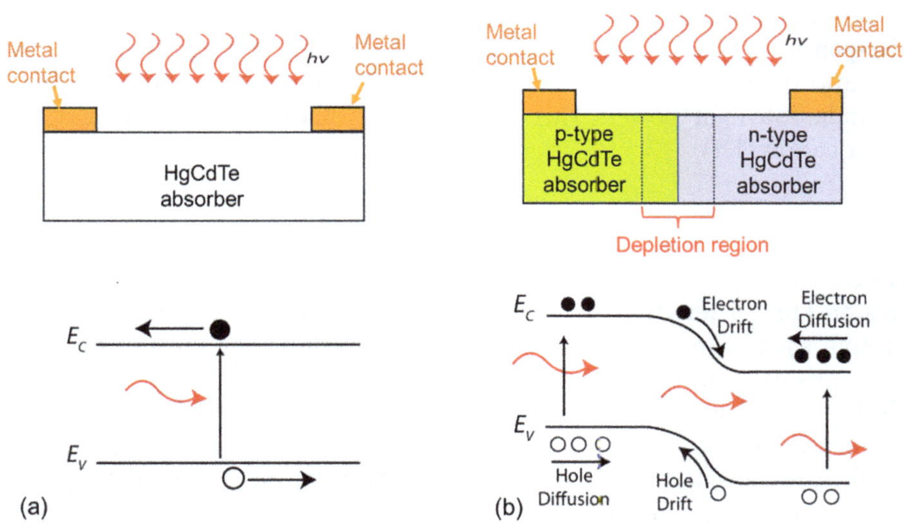

Figure 1.3. Schematic working mechanism for a HgCdTe (a) photoconductor and (b) photodiode.

and thus enhanced conductivity of the HgCdTe material. The change in conductivity is measured by supplying a constant bias current/voltage and then measuring the voltage/current signal. HgCdTe photoconductors can provide reasonably high performance, and thus have been used in various IR applications, especially single detectors and small focal plane arrays (FPAs). Single-element HgCdTe photoconductors are still widely used in spectroscopy instruments nowadays, such as Fourier transform IR spectrometers. However, because of their low impedance and high-power dissipation, HgCdTe photoconductors are only used for making small FPAs, typically <200 elements in a linear array format [10–12].

With the increased demand for high-quality imaging, which cannot be met with HgCdTe photoconductors due to their high power dissipation, large-format HgCdTe two-dimensional (2D) FPAs are needed. In contrast, photovoltaic detectors/photodiodes are p–n junction diodes operating under reverse-bias conditions, and thus have high impedance and very low power dissipation, which provides an excellent device structure for fabricating high-performance large-format 2D FPAs [10–12]. Figure 1.3(b) shows the schematic working principle of a photodiode. The IR radiation is absorbed by HgCdTe material, leading to the generation of electron-hole pairs. The generated minority carriers are free to diffuse and, upon reaching the junction depletion region, are swept across the junction by the built-in electric field, giving rise to a charge imbalance in both the n- and p-regions. Then, an external circuit can be used to measure either the resulting voltage or the photon-generated current due to this charge imbalance. Apart from very low power dissipation, photodiode detectors also have the advantages of negligible 1/f noise and easy multiplexing to a silicon readout integrated circuit (ROIC) [10–12]. Furthermore, the response of photodiodes remains linear over a significantly higher photon flux range compared to that of photoconductors. As a result, photodiodes have become the dominant device structure for IR imaging applications.

Detector figures of merit: To provide ease of comparison between detectors, some figures of merit are commonly utilized to characterize the performance of a detector.

Responsivity: The responsivity R of an IR detector is typically defined as the ratio of the average photocurrent (electrical output signal) to the incident radiation light power. This relationship is mathematically represented by equation (1.1).

$$R = \frac{I_{ph}}{P} \tag{1.1}$$

where I_{ph} is the average photo current and P is radiation power on the active region area of the detector. Note that apart from photocurrent, the electrical output signal can also be measured in photovoltage. The units of responsivity are volts per watt (V W^{-1}) or amperes per watt (A W^{-1}) [13].

Noise equivalent power: The noise equivalent power (NEP) is an indicator measured in units of watts (W) that represents the ratio of the noise current to the responsivity of photodetector. Alternatively, the NEP is the signal level that produces a signal-to-noise ratio (SNR) of 1. The NEP is expressed as equation (1.2).

$$\mathrm{NEP} = \frac{I_n}{R} \tag{1.2}$$

where I_n is the noise current, and R is the responsivity. A low NEP is ideal for all IR photodetectors since it represents the lowest light power that could be detected. Note the NEP is also quoted for a fixed reference bandwidth, which is often assumed to be 1 Hz. This "NEP per unit bandwidth" has a unit of watts per square root Hertz (W $\mathrm{Hz}^{-1/2}$) [13].

Detectivity: The detectivity D^*, in units of Jones (cm $\sqrt{\mathrm{Hz}}$/W), is a figure of merit for the sensitivity of a photodetector, defined as equation (1.3). It simplifies performance comparison by balancing the impact of active area and bandwidth on photodetectors in various geometries.

$$D^* = \frac{\sqrt{S\Delta f}}{\mathrm{NEP}} = \frac{\sqrt{S\Delta f}}{I_n}R \tag{1.3}$$

where Δf is the electrical bandwidth and S is the device active region area [13].

External quantum efficiency: The external quantum efficiency (EQE) of a photodetector is another important indicator of its ability to convert incident photons into collected photogenerated carriers that contribute to the photocurrent/photovoltage signal. It is defined as the ratio of the number of charge carriers collected by the photodetector to the number of incident photons, as expressed below:

$$\mathrm{EQE} = \frac{\text{current/charge of one electron}}{\text{total power of photons/energy of one photon}} = \frac{I_{ph}/q}{P/hv} = R\frac{hc}{\lambda q} \tag{1.4}$$

where I_{ph} is the average photocurrent, P is illumination power, q is the electron charge, h is Planck's constant, v is the light frequency, c is the speed of light in vacuum, λ is the wavelength [13]. It is evident that the EQE is highly correlated with the responsivity of the photodetector.

Response time: The response time (i.e., rise/decay or rise/reset time) is a crucial parameter for photodetectors, defined as the duration between the 10% and 90% levels of the photocurrent/photovoltage amplitude [13]. The response time of HgCdTe IR detectors is very fast which is generally in the range of several nanoseconds [11]. So, response time is not a challenge for HgCdTe IR detectors. Note that thermal detectors usually show a long response time in a range from microseconds to milliseconds (sometimes even longer) [11, 14, 15].

For FPA applications, there are a few extra figures of merit for characterizing their performance including:

Noise equivalent difference temperature (NEDT): The NEDT is defined as the temperature change of a scene required to produce a signal equal to the rms noise. In other words, it represents the minimum temperature difference which can be recognized by the FPA. The unit of NEDT is mK. Commercial HgCdTe IR FPAs typically have a NEDT <30 mK.

Pixel operability: The pixel operability of an FPA is defined as the percentage of working pixels on an FPA. A "working" pixel refers to a detector pixel that delivers signal meeting specification, while a "dead" pixel refers to a defective pixel that delivers either no signal or a signal that is out of specification. Commercial HgCdTe IR FPAs typically have a pixel operability over 99%.

Over the past several decades, various types of commercial IR detectors have been researched and developed including: Si-based bolometers, pyroelectric triglycine sulphate (TGS and DTGS), InSb photodetectors, PbSe photodetectors, HgCdTe photodetectors, quantum dot/well photodetectors, InAsSb barrier photodetectors, and type II superlattice photodetectors. Figure 1.4 shows the detectivity of various commercial IR photodetectors [11, 15]. In comparison to other IR photodetectors, HgCdTe-based IR photodetectors grown on lattice-matched CdZnTe (CZT) substrates and associated FPAs have dominated the high-performance end of IR applications due to their unbeatable performance such as a widely tuneable detection wavelength range (1–30 μm), high quantum efficiency, and very fast response rate, which will be the focus of this book.

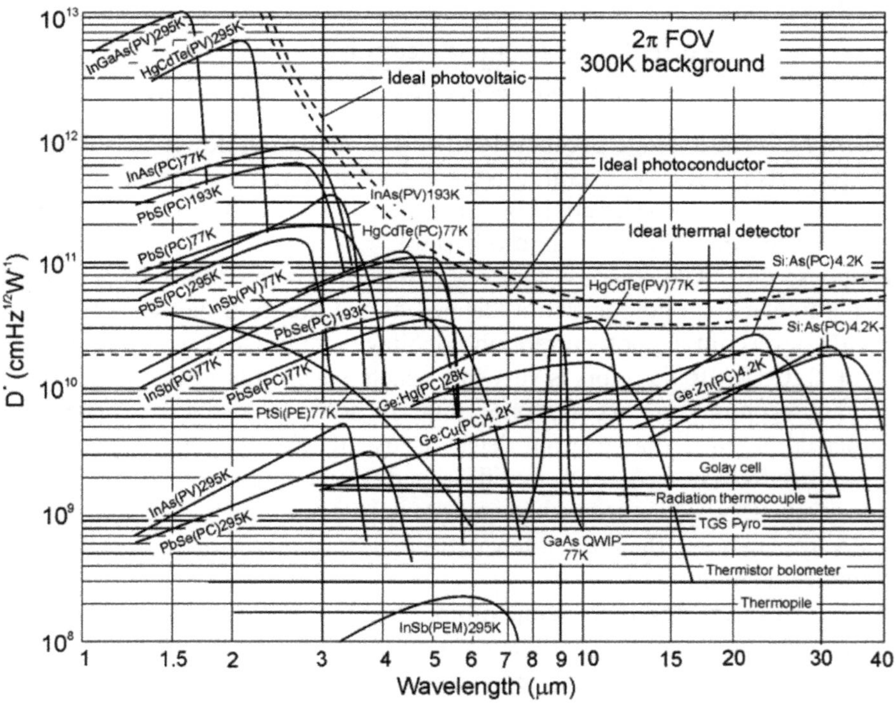

Figure 1.4. Detectivities of some typical IR detectors available on the market. Reproduced from [11]. © IOP Publishing Ltd. All rights reserved.

1.3 HgCdTe infrared detectors

Since their first discovery in 1959 by Lawson *et al* [16], HgCdTe materials develop very fast and have become the state-of-the-art materials for making high-performance IR detectors due to their favorable physical properties including (1) a tailorable energy band gap over the 1–30 μm range, (2) a large optical coefficient that enables high quantum efficiency and (3) favorable inherent recombination mechanisms that lead to high operating temperature [11]. Some key physical properties affecting the performance of HgCdTe photodetectors are as follows:

Energy band gap: An $Hg_{1-x}Cd_xTe$ semiconductor is a ternary alloy constituted by two binary alloys of HgTe and CdTe. The energy gap of this ternary alloy at 4.2 K ranges from −0.3 eV for semi-metallic HgTe, to zero at about $x = 0.15$ and then to 1.648 eV for CdTe. Figure 1.5 plots the energy band gap $E_g(x, T)$ for $Hg_{1-x}Cd_xTe$ *versus* alloy composition parameter x at temperatures 77 K and 300 K [10, 11]. The right axis of figure 1.5 also plots the cutoff wavelength $\lambda_c(x, T)$, defined as that wavelength at which the response has dropped to 50% of its peak value. There are a number of expressions for describing the energy bandgap $E_g(x, T)$ of HgCdTe [11]. The most widely used expression is as follows [11, 17]:

$$E_g = -0.302 + 1.93x - 0.81x^2 + 0.832x^3 + 5.35 \times 10^{-4}(1 - 2x)T \qquad (1.5)$$

where E_g is in the unit of electron-volt (eV) and T is in the unit of kelvin (K).

Carrier mobilities: Due to small effective masses the electron mobilities in HgCdTe are very high (typically of the order of 10^4–10^5 cm^2 V^{-1} S^{-1} for MWIR and LWIR HgCdTe at 77 K), while heavy-hole mobilities are two orders of magnitude lower (typically of the order of a few hundred cm^2 V^{-1} S^{-1} for MWIR and LWIR HgCdTe at 77 K). The electron mobility in $Hg_{1-x}Cd_xTe$ (expressed in the

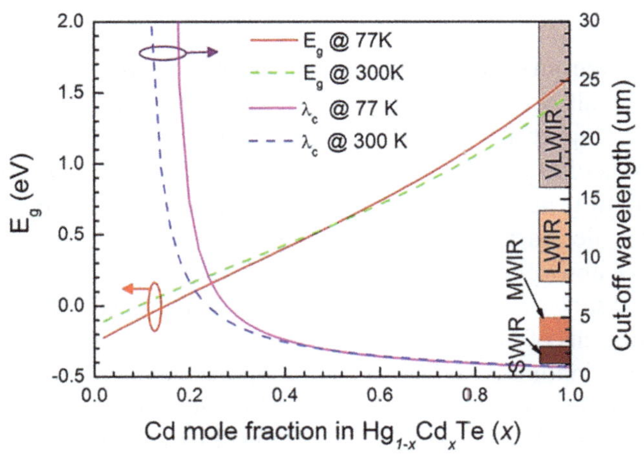

Figure 1.5. Energy band gap vs Cd composition of $Hg_{1-x}Cd_xTe$ ternary alloy at 77 K and 300 K. Reproduced from [11]. © IOP Publishing Ltd. All rights reserved.

unit of $cm^2 V^{-1} S^{-1}$), in composition range $0.2 \leqslant x \leqslant 0.6$ and temperature range $T > 50$ K, can be approximated as follows [11, 18]:

$$\mu_e = \frac{9 \times 10^8 s}{T^{2r}}, \quad \text{where } r = \left(\frac{0.2}{x}\right)^{0.6}, \quad \text{and } s = \left(\frac{0.2}{x}\right)^{7.5} \tag{1.6}$$

Note the composition range of $0.2 \leqslant x \leqslant 0.6$ covers all the main IR sub-bands: $Hg_{0.5}Cd_{0.5}Te$ for SWIR, $Hg_{0.7}Cd_{0.3}Te$ for MWIR and $Hg_{0.8}Cd_{0.2}Te$ for LWIR.

The hole mobility in HgCdTe is much lower. Typically, hole mobilities at room temperature range from 40 to 80 $cm^2 V^{-1} S^{-1}$ and the temperature dependence is relatively weak. At 77 K hole mobility is about one order of magnitude higher than that at room temperature. According to [11, 19], the 77 K hole mobility reduces as the acceptor concentration increases and the following empirical expression can be used to approximate it in the composition range of 0.20–0.30:

$$\mu_h = \mu_0 \left[1 + \left(\frac{p}{1.8 \times 10^{17}}\right)^2\right]^{-1/4} \tag{1.7}$$

where $\mu_0 = 440$ $cm^2 V^{-1} S^{-1}$, and p is hole concentration.

For modeling IR photodetectors, the hole mobility is usually calculated assuming that the electron-to-hole mobility ratio $b = \mu_e/\mu_h$ is constant and equal to 100 [11]. It should be noted that both the electron and hole mobilities are affected by the defects and impurities in the HgCdTe materials. Therefore, the above empirical expressions only provide a guidance, and the actual mobilities should be measured with Hall measurements on the HgCdTe materials.

Optical properties: One of the critical optical properties for photodetector materials is the optical absorption coefficient *a*. There are a number of empirical expressions for calculating the intrinsic optical absorption coefficient of HgCdTe materials [11]. Among them, the most widely used is that proposed by Chu *et al* [11, 20]:

$$\alpha = \alpha_g \exp\left[\beta(E - E_g)\right]^{1/2} \tag{1.8}$$

where a_g is the absorption coefficient at the band gap energy E_g, and the parameter β depends on the alloy composition and temperature: $\beta(x, T) = -1 + 0.083\, T + (21-0.13\, T) \times x$. Figure 1.6 plots the optical absorption coefficient data for several alloy compositions of $Hg_{1-x}Cd_xTe$ versus wavelength [11, 21]. It is observed that the absorption strength generally decreases as the material energy gap becomes smaller, due to both the decrease in the conduction band effective mass and the $\lambda^{-1/2}$ dependence of the absorption coefficient on wavelength λ. It should also be mentioned that the optical absorption coefficient is also affected by the HgCdTe material quality such as concentrations of native defects and impurities, non-uniform composition and doping, thickness inhomogeneities of samples, mechanical strains, and different surface treatments [11], which should be taken into account in real experiments.

As reported before [11, 22], the dielectric constants were not a linear function of Cd composition x and temperature dependence was not observed in the experiments. These dependences can be described by the following expressions:

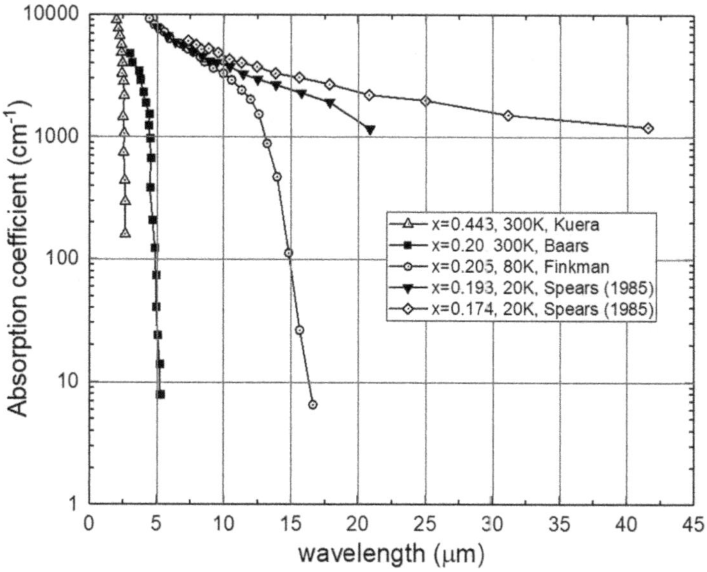

Figure 1.6. Optical absorption coefficient data for several $Hg_{1-x}Cd_xTe$ alloy compositions, for photon energies near the fundamental absorption edge. Reproduced from [11]. © IOP Publishing Ltd. All rights reserved.

$$\varepsilon_\infty = 15.2 - 15.6x + 8.2\,x^2 \tag{1.9}$$

$$\varepsilon_0 = 20.5 - 15.6x + 5.7\,x^2 \tag{1.10}$$

1.4 Current status and challenges of HgCdTe infrared detectors

After several decades of development, significant progress has been achieved in this strategically important research area of HgCdTe IR detectors. So far, high-performance HgCdTe IR detectors and their FPAs are available for various IR sensing and imaging applications. Next, we will summarize the current status and challenges of HgCdTe IR detectors.

Material growth: In general, there are three main types of epitaxial growth techniques for HgCdTe. They are: liquid phase epitaxy (LPE), metal–organic chemical vapor deposition (MOCVD) and molecular beam epitaxy (MBE). LPE is the most traditional and mature epitaxy technology and has been used widely for industrial production of HgCdTe for several decades. In order to achieve higher crystal quality and growth efficiency, HgCdTe materials are usually grown on (111) CdZnTe (4% Zn) substrates, which are lattice-matched to HgCdTe. The HgCdTe materials obtained via LPE show high crystalline quality, typically demonstrating x-ray diffraction full-width-half-maximum (XRD FWHM) of 25–40 arcsec, an etch pit density (EPD) of 1×10^4–1×10^5 cm^{-2}, and a residual doping of $<1 \times 10^{15}$ cm^{-3} [23, 24]. The LPE method has been successfully applied to growing HgCdTe detectors in various spectral detection bands, including SWIR, MWIR, LWIR, and VLWIR. It is expected that LPE will continue to be an important growth method for HgCdTe, especially LWIR HgCdTe. However, this method doesn't allow the epitaxial growth of complex detector structures, such as multiband detectors, barrier detectors, etc. As a result, significant attention has shifted to MOCVD and MBE growth of HgCdTe.

In comparison to LPE, MOCVD and MBE allow the growth of larger area wafers and more complex device structures with good lateral homogeneity and abrupt and complex composition and doping profiles, which are essential requirements for the development of advanced HgCdTe detectors and FPAs. Specifically, MOCVD and MBE enjoy the flexibility in engineering the material and device parameters during growth, such as bandgap energies, incorporation of heterostructures and multiple layers, control of doping profiles, use of alternative substrates, formation of passivation layers, in-situ annealing, etc. In particular, they have been an indispensable tool for developing dual-band IR detectors that integrate two detectors onto a single pixel, which is a complex device structure. Nowadays, the material quality obtained with MOCVD and MBE is comparable to that routinely obtained with LPE. MOCVD growth of HgCdTe is mainly conducted in the United Kingdom (Selex) and Poland (Vigo), while MBE growth is undertaken in many other countries, including the United States, France, Russia, Germany, Israel, China, Norway, South Korea, and Australia. Because of the higher popularity of MBE growth for HgCdTe materials, this book will mainly focus on the MBE growth of HgCdTe material for IR detector applications.

Device architectures: With the progress of material growth techniques over the past several decades, HgCdTe detector architectures have also developed from the initial simple photoconductor structure to a standard p–n junction photodiode structure and, most recently, to more complex detector structures. Among them, the p–n junction photovoltaic diode structure has attracted more attention and become the mainstream technology. In general, there are two groups of p–n junction photodiodes: one is an n-on-p junction diode, the other is a p-on-n junction diode, where the latter device has been demonstrated to have higher device performance: lower dark current and thus higher operating temperature [10, 25]. Both n-on-p and p-on-n junction diode structures have been successfully applied for fabricating HgCdTe FPAs with large format size. Recently, other improved p–n junction photovoltaic diode structures have also been studied to enhance device performance, especially multiband detection and operating temperature. For example, back-to-back p–n junction diode structures were applied to achieve dual-band detection for HgCdTe IR detectors [26], and fully depleted p–n junction diode structures were used to achieve high operating temperature for HgCdTe IR detectors [27].

Device performance: After developing over several decades, current HgCdTe IR detectors present remarkable device performance. Some key features are summarized as below: (1) detectivity: as shown in figure 1.4 HgCdTe IR detectors present very high detectivity with typical values of mid-10^{12} cm $\sqrt{\mathrm{Hz}}$ W^{-1} for SWIR, mid-10^{11} cm $\sqrt{\mathrm{Hz}}$ W^{-1} for MWIR, mid-10^{10} cm $\sqrt{\mathrm{Hz}}$ W^{-1} for LWIR; (2) array format: commercial HgCdTe detectors nowadays have various array format sizes ranging from 320 × 256, to 640 × 512, to 1 K × 1 K, to 2 K × 2 K, and 4 K × 4 K maximum. Large array formats (≥1 K × 1 K) are mainly used for astronomy study such as Teledyne HgCdTe IR detectors [28]; (3) NEDT: commercial HgCdTe FPAs typically have a small NEDT (<30 mK). Some HgCdTe FPAs have an NEDT less than 20 mK [29]; (4) operating temperature: HgCdTe IR detectors are typically required to work at low temperature (e.g., 77 K). However, HgCdTe IR detectors with a special design for higher operating temperature can operate up to 150 K for MWIR and 110 K for LWIR [10, 29]. Most recently, Teledyne proposed a unique fully depleted diode structure and successfully enhanced the operating temperature of MWIR HgCdTe detector up to 215 K [27, 30]; (5) multiband detection: in addition to single band detectors, dual band HgCdTe detectors are also available from some manufacturers (Raytheon, AIM, CEA–Leti, etc) [26, 31–34], including SWIR/MWIR and MWIR/LWIR.

Challenges for current HgCdTe infrared detectors: With the increased demand of various IR industry sectors, especially military applications, more enhanced features are required for the future development of IR detectors, which cannot be met with the current state-of-the-art IR detectors. Generally, the new features include [10–12, 35]: (1) high-performance, high-resolution cooled imagers having multicolor bands, (2) medium- to high-performance uncooled imagers, (3) larger array format size with smaller area pixels, and (4) lower cost. IR detectors with these new features are usually referred to as 'third-generation' IR detectors. Although some other competing IR technologies have been developed, HgCdTe IR detectors will continue to dominate the high-performance end of the IR market in the foreseeable future. Therefore, HgCdTe will still be the material of choice for developing 'third-generation' IR detectors and their FPAs. As a result, this book will focus on addressing the two features of lower cost and larger array format size, and discussing various approaches to achieve HgCdTe IR detectors with these two features.

1.5 How to achieve lower cost and larger array format size for HgCdTe detectors

Cost, format size and yield issues [10]: Two of the requirements for next-generation HgCdTe IR detectors are lower cost and larger array format size. The single dominant factor that impacts on the high cost of HgCdTe-based FPAs is very low yield, particularly due to the absence of large-area, high-quality, lattice-matched substrates. It is generally well-known, although not widely acknowledged, that the very low pixel-yield on an FPA, and the corresponding low FPA yield on a wafer, is dominated by defects resulting from the relatively low quality buffer/substrate starting materials. For current state-of-the-art HgCdTe IR technologies, high-performance HgCdTe IR detectors are generally grown on lattice-matched CdZnTe substrates. However, for lattice-matched CdZnTe substrates, there are no suppliers of epi-ready substrates and the defect density of the substrates, in terms of EPD, is one to two orders of magnitude higher than that of other commercially available epi-ready substrates (for example, ~low 10^3 cm^{-2} for GaAs). During epitaxial growth, these substrate defects will evolve into defects in the HgCdTe epi-layers (such as threading dislocations, point defects, etc), which results in 'dead' pixels in the FPAs due to the excessive dark current that results whenever the pixel active area coincides with an electrically active defect. Note that a 'dead' pixel is defined as a defective pixel that delivers either no signal or a signal that is out of specification. Thus, a high defect density leads to very low yield and, as a result, very high cost for current-generation HgCdTe FPAs. Apart from the cost issue, a high defect density also limits the array size of an FPA, since the probability of success in fabricating a functioning FPA decreases dramatically with increasing area.

In addition to relatively lower crystal quality, CdZnTe substrates also suffer two other limitations, these being high cost and small wafer size. For example, a piece of 1 cm × 1 cm CdZnTe substrate costs approximately \$300, and the maximum wafer size for commercially available CdZnTe substrates is currently limited to 8 cm × 8 cm. Another disadvantage of CdZnTe substrates is that there are very few commercial suppliers that can provide high-quality CdZnTe substrates. Therefore, there is a strong incentive to replace CdZnTe substrates with alternative high-quality, large-area and low-cost substrates in order to enhance FPA yield, increase wafer size and, thus increase the number of fully-functional FPAs per wafer in order to reduce the overall cost.

Alternative substrates [10]: Over the past two decades, several alternative substrates have been studied to replace CdZnTe for growing HgCdTe materials, such as Si, Ge, and GaAs [36–41]. All these substrates have the features of high crystal quality (in terms of etch pit density (EPD)), large wafer size, lower cost, and ready availability. However, all these alternative substrates present a very large lattice and CTE (coefficient of thermal expansion) mismatch with HgCdTe materials. Figure 1.7 shows the lattice and CTE mismatch between HgCdTe and the substrates being discussed [41]. The large lattice and CTE mismatches between HgCdTe and alternative substrates lead to a high defect density (typically mid-10^6 cm^{-2} to low-10^7 cm^{-2}) in the MBE-grown HgCdTe epilayers, which not only

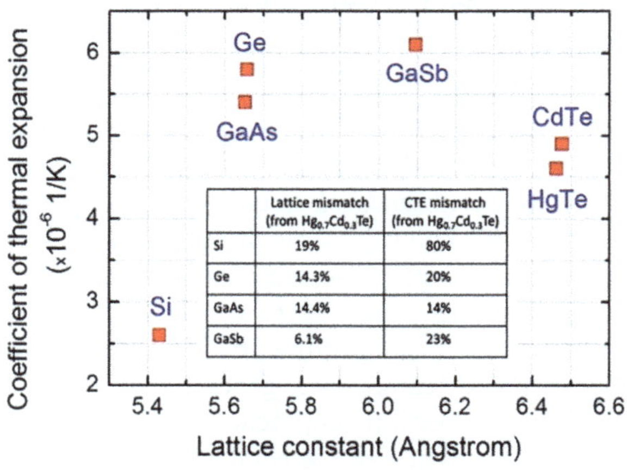

Figure 1.7. Room temperature lattice constant and CTEs of several semiconductors used as substrates for HgCdTe MBE growth. The inset shows the lattice and CTE mismatch between four potential alternative substrates and HgCdTe. Reproduced from [41] with permission from Springer Nature.

reduces the carrier mobility and lifetime (and thus degrades detector performance), but also leads to a high density of defective/dead pixels in the FPA. Although such high dislocation densities can be tolerated by MWIR HgCdTe detectors, they significantly degrade the performance of LWIR HgCdTe detectors because LWIR devices are more sensitive to the defects due to their narrower bandgap. To achieve high performance in LWIR HgCdTe FPAs, the dislocation density in HgCdTe needs to be less than 5×10^5 cm^{-2} [36], which renders current Si, Ge and GaAs alternative substrate technologies unsuitable for growing LWIR HgCdTe epitaxial materials. Therefore, the key to the application of alternative substrates for fabricating high-performance HgCdTe IR detectors is to develop either new alternative substrate technologies or new approaches to improve the existing alternative substrate technologies.

This book will start with the fundamental mechanism of lattice-mismatched epitaxial growth in order to better understand the physics/chemistry behind lattice-mismatched epitaxial growth, and the generation, propagation and evolution of misfit/threading defects during the growth; then discuss various material platforms for growing CdTe and HgCdTe layers on lattice-mismatched substrates and various approaches to reduce/suppress the generation of misfit and threading dislocations, control the propagation and evolution of misfit/threading dislocations, and block/annihilate misfit/threading dislocations; then discuss HgCdTe IR detectors grown on lattice-mismatched substrates; then discuss other alternative IR material systems (e.g., HgCdSe) to achieve high-performance IR detectors with lower cost and larger arrays, and, finally, present a summary of this book and outlook for the future development of lattice-mismatched epitaxy for fabricating high-performance HgCdTe IR materials and detectors.

References

[1] https://columbia.edu/~vjd1/electromag_spectrum.htm

[2] https://en.wikipedia.org/wiki/Infrared

[3] Hudson R 1969 *Infrared System Engineering* (New York: Wiley)

[4] https://dyt-ir.com/news/65-unique-uses-of-infrared-thermal-imaging-cameras

[5] Beletic J W, Blank R, Gulbransen D, Lee D, Loose M, Piquette E C, Sprafke T, Tennant W E, Zandian M and Zino J 2008 Teledyne imaging sensors: infrared imaging technologies for astronomy and civil space *Proc. SPIE* **7021** 70210H

[6] https://en.wikipedia.org/wiki/Bolometer

[7] https://en.wikipedia.org/wiki/Pyroelectricity

[8] Sewell R H 2005 Investigation of mercury cadmium telluride heterostructures grown by molecular beam epitaxy *PhD Thesis* The University of Western Australia

[9] Manh Nguyen T H 2005 A photovoltaic detector technology based on plasma-induced p-to-n type conversion of long wavelength infrared HgCdTe *PhD Thesis* The University of Western Australia

[10] Lei W, Antoszewski J and Faraone L 2015 Progress, challenges and opportunities for HgCdTe infrared materials and detectors *Appl. Phys. Rev.* **2** 041303

[11] Rogalski A 2005 HgCdTe infrared detector material: history, status and outlook *Rep. Prog. Phys.* **68** 2267

[12] Norton P 2002 HgCdTe infrared detectors *Opto-Electron. Rev.* **10** 159
[13] Wang H 2025 Controlled growth, characterization, and applications of low-dimensional bismuth chalcogenides *PhD Thesis* The University of Western Australia
[14] Stillman G E 2002 Optoelectronics *Reference Data for Engineers* 9th edn ed W M Middleton and M E Van Valkenburg (Amsterdam: Elsevier) pp 21–1–21–31
[15] Rogalski A 2020 *Infrared and Terahertz Detectors* 3rd edn (Boca Raton, FL: CRC Press) pp 54–8
[16] Lawson W D, Nielson S, Putley E H and Young A S 1959 Preparation and properties of HgTe and mixed crystals of HgTe–CdTe *J. Phys. Chem. Solids* **9** 325
[17] Hansen G L, Schmit J L and Casselman T N 1982 Energy gap versus alloy composition and temperature in $Hg_{1-x}Cd_xTe$ *J. Appl. Phys.* **53** 7099
[18] Rosbeck J P, Star R E, Price S L and Riley K J 1982 Background and temperature dependent current–voltage characteristics of $Hg_{1-x}Cd_xTe$ photodiodes *J. Appl. Phys.* **53** 6430
[19] Dennis P N J, Elliott C T and Jones C L 1982 A method for routine characterization of the hole concentration in p-type cadmium mercury telluride *Infrared Phys.* **22** 167
[20] Chu J, Li B, Liu K and Tang D 1994 Empirical rule of intrinsic absorption spectroscopy in $Hg_{1-x}Cd_xTe$ *J. Appl. Phys.* **75** 1234
[21] Reine M B 2004 *Fundamental Properties of Mercury Cadmium Telluride Encyclopedia of Modern Optics* (London: Academic)
[22] Dornhaus R, Nimtz G and Schlicht B 1983 *Narrow-Gap Semiconductors* (Berlin: Springer)
[23] Gravrand O and Destefanis G 2013 Recent progress for HgCdTe quantum detection in France *Infrared Phys. Technol.* **59** 163
[24] Gravrand O *et al* 2013 Issues in HgCdTe research and expected progress in infrared detector fabrication *J. Electron. Mater.* **42** 3349
[25] Rogalski A and Larkowski W 1985 Comparison of photodiodes for the 3–5.5 μm and 8–12 μm spectral regions *Electron. Technol.* **18** 55
[26] Ballet P *et al* 2004 Dual-band infrared detectors made on high-quality HgCdTe epilayers grown by molecular beam epitaxy on CdZnTe or CdTe/Ge substrates *J. Electron. Mater.* **33** 667
[27] Lee D *et al* 2020 Law 19: the ultimate photodiode performance metric *Proc. SPIE* **11407** 114070X
[28] https://teledynespaceimaging.com/en-us/Products_/Pages/infrared-hgcdte-chroma-d.aspx
[29] Mollard L *et al* 2014 p-on-n HgCdTe infrared focal-plane arrays: from short-wave to very-long-wave infrared *J. Electron. Mater.* **43** 802
[30] Lee D, Carmody M, Piquette E, Dreiske P, Chen A, Yulius A, Edwall D, Bhargava S, Zandian M and Tennant W E 2016 High-operating temperature HgCdTe: a vision for the near future *J. Electron. Mater.* **45** 4587–95
[31] Destefanis G, Baylet J, Ballet P, Castelein P, Rothan F, Gravrand O, Rothman J, Chamonal J P and Million A 2007 Status of HgCdTe bicolor and dual-band infrared focal arrays at LETI *J. Electron. Mater.* **36** 1031
[32] Ballet P *et al* 2005 Demonstration of a 25μm pitch TV/4 dual-band HgCdTe infrared focal plane array with spatial coherence *Proc. SPIE* **5957** 595703
[33] Smith E P G, Venzor G M, Gallagher A M, Reddy M, Peterson J M, Lofgreen D D and Randolph J E 2011 Large-format HgCdTe dual-band long-wavelength infrared focal-plane arrays *J. Electron. Mater.* **40** 1630

[34] https://aim-ir.com/en/applications-products/security/modules/3rd-gen-ir-modules/hipir-320mm.html
[35] Rogalski A, Antoszewski J and Faraone L 2009 Third-generation infrared photodetector arrays *J. Appl. Phys.* **105** 091101
[36] Reddy M *et al* 2011 Molecular beam epitaxy growth of HgCdTe on large-area Si and CdZnTe substrates *J. Electron. Mater.* **40** 1706
[37] Carmody M *et al* 2012 Recent progress in MBE growth of CdTe and HgCdTe on (211)B GaAs substrates *J. Electron. Mater.* **41** 2719
[38] Wenisch J, Eich D, Lutz H, Schallenberg T, Wollrab R and Ziegler J 2012 MBE growth of MCT on GaAs substrates at AIM *J. Electron. Mater.* **41** 2828
[39] He L *et al* 2007 MBE HgCdTe on Si and GaAs substrates *J. Cryst. Growth* **301–302** 268
[40] Zanatta J P *et al* 2006 Molecular beam epitaxy growth of HgCdTe on Ge for third-generation infrared detectors *J. Electron. Mater.* **35** 1231
[41] Lei W, Gu R J, Antoszewski J, Dell J and Faraone L 2014 GaSb: a new alternative substrate for epitaxial growth of HgCdTe *J. Electron. Mater.* **43** 2788

Chapter 2

General growth mechanism of heteroepitaxy

In chapter 1, we presented a brief introduction of IR spectral band, IR sensing and imaging, and the core electronic component of IR application system—IR detectors. Among various IR detectors, HgCdTe IR detectors have dominated the high-performance end of the IR application market for decades. However, as discussed in chapter 1, HgCdTe IR detectors suffer a major disadvantage—higher cost and smaller array format size, which is mainly caused by their lattice-matched CdZnTe substrates due to the higher cost, lower crystal quality, smaller wafer size, and limited commercial suppliers. The future development of IR applications requires the next generation of high-performance IR detectors to have the favorable features of lower cost and larger array format size in order to broaden their industry applications (especially civilian applications). As a result, it is critical to achieve high-quality HgCdTe materials on various alternative substrates of lower cost, higher crystal quality, larger wafer size, and wider commercial availability to enhance the device yield, lower the cost, and increase the array format size. Before achieving high-quality HgCdTe materials on alternative substrates, it is essential to have a deep understanding about the growth mechanisms of hetero-epitaxial growth, especially lattice-mismatched one. Thus, this is what chapter 2 will focus on. Special attention will be given to various general techniques for suppressing defect generation and control and annihilation of defects within the epilayers with the ultimate goal of having high-quality HgCdTe on alternative substrates for making high-performance HgCdTe IR detectors.

2.1 Introduction to semiconductor heterostructures and heteroepitaxy

Semiconductor heterostructures are structures composed of two or more semi-conductor materials with different chemical compositions, creating interfaces where

doi:10.1088/978-0-7503-3443-3ch2 © IOP Publishing Ltd 2025. All rights, including for text and data mining (TDM), artificial intelligence (AI) training, and similar technologies, are reserved.

the material properties change. These interfaces are crucial for tailoring the electronic and optical properties and thus they enable improved performance and new functionalities of semiconductor devices. Nowadays, semiconductor heterostructures have become the building blocks for modern electronic and optoelectronic devices. For example, semiconductor heterostructures are widely used for fabricating high-performance semiconductor lasers and photodetectors, especially quantum well and superlattice devices. One of the most elegant applications of semiconductor heterostructures is quantum cascade devices, including quantum cascade lasers (QCLs) and detectors (QCDs), the active regions of which consist of multiperiods of quantum well or superlattice heterostructures. By choosing the material and thickness of quantum wells/superlattices, these QCLs can emit at wavelengths ranging from SWIR to LWIR, such as InAs/AlSb SWIR QCLs [1], MWIR InGaAs/InAlAs QCLs [2], and LWIR InGaAs/InAlAs QCLs [3]. This fully demonstrates the importance of heterostructures for modern electronic and optoelectronic devices.

How are these semiconductor heterostructures fabricated? Typically, they are formed by epitaxial growth of one material on another. Epitaxial growth refers to the ordered growth of a crystalline layer on a substrate, with the layer replicating the substrate's crystallographic structure. This precise atomic arrangement is critical for producing high-quality semiconductor materials for use in high-performance semiconductor devices. With advances in epitaxial methods such as chemical vapor deposition (CVD) and MBE, various semiconductor heterostructures can be achieved, including simple heterojunction, quantum wells, and superlattices, which enable the development of devices with superior performance. The main benefits for constituting heterostructures include:

Bandgap engineering: By combining materials with different bandgaps and properties, heterostructures can be tailored to achieve specific electronic and optical characteristics. Examples include GaAs/AlGaAs quantum wells [4].

Device integration: Heteroepitaxy enables integration of disparate materials into complex architectures, reducing system size and cost while enhancing device performance.

Substrate engineering: Many promising semiconductors are not available as large, high-quality wafers such as CdZnTe substrates for growing HgCdTe materials [5]. Heteroepitaxy on alternative substrates is necessary to enable scalable fabrication such as heteroepitaxy of HgCdTe on Si, Ge, and GaAs substrates [5].

Therefore, it is essential to undertake high-quality heteroepitaxial growth by depositing one semiconductor material layer on another in order to make high-quality semiconductor heterostructures. Depending on the lattice constant of the two materials constituting the heterojunction, heteroepitaxial growth can be categorized into two types: lattice-matched heteroepitaxy and lattice-mismatched heteroepitaxy. Both are critical for fabricating electronic and optoelectronic devices.

Lattice-matched heteroepitaxy: Lattice-matched heteroepitaxy involves growing a layer with nearly identical lattice parameters to the substrate, resulting in minimal defects and high-quality films.

Lattice-mismatched heteroepitaxy: Lattice-mismatched heteroepitaxy involves growing materials with differing lattice parameters and/or compositions, which introduces strain and potential defects but also enables enhanced functionalities.

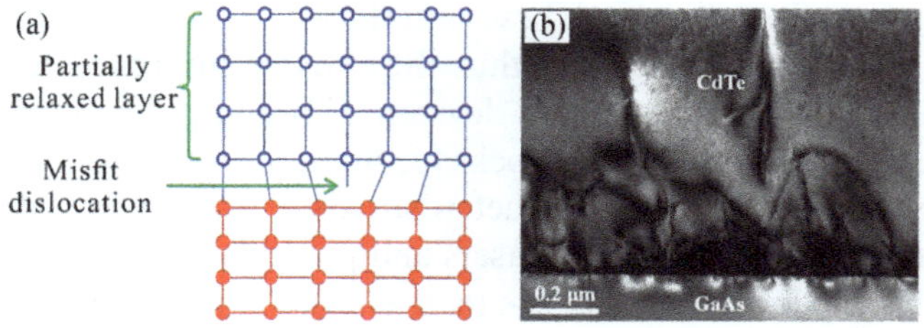

Figure 2.1. (a) Schematic formation of misfit dislocations due to lattice mismatch, and (b) cross-sectional transmission electron microscopy (TEM) image of CdTe epitaxial layers grown on GaAs showing the vertical propagation of threading dislocations. Panel (b) was reproduced from [7] with permission from Springer Nature.

It should be noted that each material, by nature, has its own unique physical properties, in particular crystal lattice constant, which is usually not well matched across the heterostructure interface. As sketched in figure 2.1(a), this lattice mismatch between the two materials results in accumulation of misfit strain (elastic) with increasing thickness of the top layer material during heteroepitaxy [6] and, once the top layer exceeds its critical thickness the accumulated strain will be relaxed to stabilize the system by forming misfit dislocations in the vicinity of the interface. As demonstrated in figure 2.1(b) [7], some of these dislocations will propagate vertically (along the growth direction) to form threading dislocations which will result in defect energy levels, as well as carrier scattering and recombination centers in the epitaxial layers. Such defects can have a disastrous impact on device yield, as well as material physical properties including carrier mobility, carrier lifetime, background doping concentration, etc, which will degrade the performance of functioning devices. Therefore, the success of heterostructure devices strongly relies on the material quality of the grown heterostructures: in particular the dislocation density in the epitaxial layers must be lower than a certain level in order to achieve the desirable device performance. Because of this, current high-performance heterostructure devices are severely limited since they are reliant on lattice-matched material systems in order to avoid the generation of misfit and threading dislocations, such as InGaAs/InAlAs QCLs on lattice-matched InP substrates [8, 9].

With increasing industry applications, significant attention has been devoted to heteroepitaxy with large lattice mismatch, such as III–V semiconductors (GaAs, InP, GaSb, InSb) on Si, as well as II–VI semiconductors on Si, Ge and III–V substrates. This is due to the fact that the number of lattice matched heterostructure material systems is very limited, and cannot meet the increasing performance requirements of future industry applications. For example, III–V semiconductors such as GaAs and InP enjoy the favourable properties of direct bandgap, and high carrier mobility, and thus are widely used for fabricating optoelectronic devices. In comparison, apart from being of lower cost, larger wafer size, and higher mechanical robustness, Si enjoys considerable advantages in terms of manufacturing due to its long history and huge microelectronic manufacturing base currently in place. However, Si is an indirect bandgap semiconductor, and thus not suitable for fabricating high performance optoelectronic devices. Therefore, heteroepitaxy of III–V semiconductors such as GaAs and InP on Si will leverage the advantages of both III–V semiconductors and Si, as well as achieve monolithic integration of electronic and photonic devices. The principal challenge for these heteroepitaxial processes relates to the large lattice mismatch and thus high density of threading dislocations generated in the epitaxial material, which significantly deteriorates the ultimate device performance and yield.

Although significant progress has been made in high-quality heteroepitaxy, the performance of devices based on lattice-mismatched heteroepitaxy is still much lower than that of their counterparts grown on lattice-matched substrates due to the high dislocation density generated during heteroepitaxy. A typical example is heteroepitaxy of the strategically important II–VI semiconductor HgCdTe for the fabrication of IR detectors on lattice mismatched Si, Ge, and GaAs substrates with the aim of lowering the production cost, increasing the detector FPA format size, enhancing system robustness, and achieving potential monolithic integration with microelectronic read-out circuits. Although the dislocation density in HgCdTe epilayers grown on Si, Ge, and GaAs has been reduced to the level of mid-10^6 to low-10^7 cm^{-2} after two decades of effort [10–12], it is still almost two orders of magnitude higher than that in the counterpart grown on lattice matched CdZnTe substrates [10]. This seriously degrades the device performance of fabricated IR detectors, especially LWIR detectors. To fabricate high-performance HgCdTe LWIR detectors, the dislocation density must be controlled below the level of 5×10^5 cm^{-2} [5], which requires a significant reduction in the density of dislocations from what is currently being achieved on lattice-mismatched substrates. That is what this book focuses on—how to achieve high-quality HgCdTe materials on lattice-mismatched alternative substrates.

In addition, future semiconductor devices are required to have some novel characteristics, such as large-scale, cost-effective monolithically integrated devices, flexible devices, etc, which will require the capability to be able to grow large-area, free-standing, lattice-matched, and even lattice-mismatched epitaxial films. This requires a new epitaxial growth mechanism such as the emerging Van der Waals (vdW) heteroepitaxy [13], which bypasses traditional lattice-matching constraints by using weak interlayer forces, expanding possibilities for novel material growth and future demanding applications.

To achieve high-quality heteroepitaxial growth and thus high-quality heterostructures, we need to have a sound understanding of the related mechanisms of semiconductor heteroepitaxial growth. In this chapter, section 2.2 will briefly introduce the growth mechanism of lattice-matched heteroepitaxy; section 2.3 will discuss the growth mechanism of lattice-mismatched heteroepitaxy and defect engineering approaches; while section 2.4 will explore lattice-mismatched heterostructure growth based on novel Van der Waals epitaxy.

2.2 Growth mechanism of lattice-matched heteroepitaxy

Generally, heteroepitaxial growth starts from substrates by growing one layer on another. Basically, a substrate essentially acts as a seeding crystal, and the post-grown film locks into the substrate's crystallographic orientation by using suitable growth techniques such as MBE or MOCVD. When new atoms come onto the substrate surface, they can absorb, desorb, diffuse, and join or nucleate with another atom to form little islands. Islands can also grow, migrate, or evaporate, and then at some critical size, they begin to grow outward and merge with other islands by locking to the surface potential wells, leading to crystallographic mimicry for the added layer. Therefore, substrates with high crystal quality and low defect density will be the key to achieve high-quality heteroepitaxial growth. Fortunately, substrate technology has advanced significantly over the past several decades, and nowadays various high-quality substrates are commercially available such as SiC, Sapphire (Al_2O_3), GaN, Si, Ge, GaAs, InP, InAs, GaSb, InSb, and CdZnTe. Note that CdZnTe substrates are very special, have very limited commercial availability, and are not epi-ready substrates [5].

For lattice-matched heteroepitaxial growth, one material will be deposited on another with the same or nearly matched lattice constant. Because all the layers in the lattice-matched heterostructures are lattice-constant matched, no misfit dislocations will be generated, leading to the epitaxial growth of thin film materials with high crystal quality and minimum defect density. For this type of lattice-matched heteroepitaxial growth, the crystal quality and defect density of the epitaxial layers are basically determined by those of substrates as they inherit whatever exists in the substrates. So far, current high-performance semiconductor devices are all based on these lattice-matched heterostructures. Some examples include: III–V GaAs substrate-based GaAs/AlGaAs heterostructures are nearly lattice-matched material systems that are widely used for SWIR LEDs, lasers, high-speed electronics such as field-effect transistors (FETs), and high-electron-mobility transistors (HEMTs) [14, 15]. III–V GaSb substrate-based InAs/GaSb type-II superlattice is an emerging nearly lattice-matched material for IR detection [16]. II–VI HgTe and CdTe also have similar lattice constants and the lattice-matched HgCdTe growth on CdZnTe substrates has led to state-of-the-art IR imaging FPA technologies [17].

Generally, the lattice-matched heteroepitaxy process involves the following steps:

Cleaning and preparation of substrates: Substrates are usually covered with a thin layer of native oxide due to exposure to air, which needs to be removed for subsequent epitaxial growth. The oxide layer is usually chemically etched off and/or thermally desorbed by heating. In addition to removing native oxide, substrates are required to be cleaned thoroughly to remove any impurities or surface contaminants that can negatively affect the growth of subsequent epitaxial layers.

Epitaxy of buffer and device layers: After oxide removal, substrates usually present a rough surface, which is not suitable for growing subsequent device layers directly. As a result, a homogenous or lattice-matched epitaxial growth of a thin buffer layer is needed to flatten the surface for the subsequent growth of high-quality device layers.

Overall, lattice-matched heteroepitaxy provides a powerful tool for the design and fabrication of advanced electronic and optoelectronic devices, and it continues to play an important role in the development of new device technologies.

2.3 Growth mechanisms of lattice-mismatched heteroepitaxy

In contrast to lattice-matched, lattice-mismatched heteroepitaxy is the growth of a crystalline material on a substrate with a different crystalline structure in terms of lattice constant and crystal phase, resulting in a strain between epitaxial layer and substrate. Another concern is their mismatch in CTE (coefficient of thermal expansion), which can also result in strain. The strain can be accommodated by different mechanisms, including the formation of misfit and threading dislocations. To achieve high-quality lattice-mismatched heteroepitaxy, it is essential to have a good understanding of its related growth mechanisms.

2.3.1 Growth modes

For heteroepitaxy, there are three main growth modes that can occur as the deposited material thickness approaches and exceeds the critical thickness: layer-by-layer growth, three-dimensional island growth, and layer-plus-island growth [18]. Figure 2.2 shows the schematic diagrams for the three growth modes. The growth mode that occurs in heteroepitaxy depends on the lattice mismatch and elastic properties of substrate and deposited material. Layer-by-layer growth mode is desirable for most applications as it results in high-quality thin films, but it is only possible when lattice mismatch is low and critical thickness is high. If lattice mismatch is high, three-dimensional island growth mode is more likely to occur, and buffer layers may be necessary to reduce the defect density.

Layer-by-layer growth: Layer-by-layer growth is also called Frank–van der Merwe growth. In this growth mode, the deposited material grows in a layer-by-layer fashion without forming any islands or dislocations. This mode occurs when the deposited material thickness is much less than the critical thickness, and the elastic energy can be fully accommodated by the substrate. This growth mode results in high-quality thin films with excellent surface smoothness and low defect density.

Stranski–Krastanov (SK) or layer-plus-island growth: The film initially grows layer-by-layer, wetting the substrate surface. However, at a certain thickness (critical thickness), it becomes energetically more favorable to form three-dimensional islands on top of the wetting layer. This mode results in a combination of a thin film and islands. This growth mode leads to the formation of self-assembly nano-structures such as quantum dots, quantum wires, and quantum rings [19].

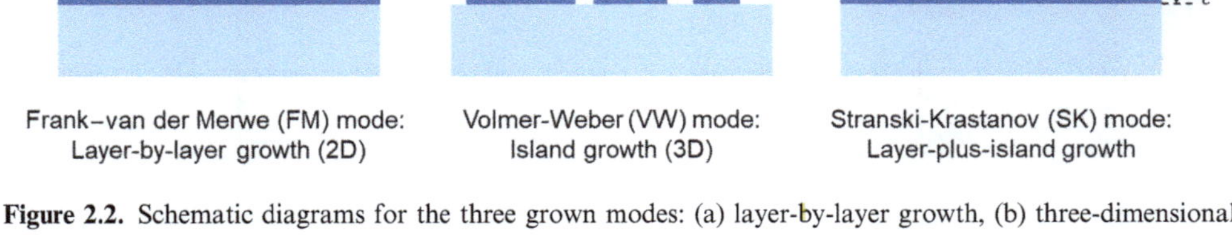

Figure 2.2. Schematic diagrams for the three grown modes: (a) layer-by-layer growth, (b) three-dimensional island growth, and (c) layer-plus-island growth. Reproduced from [18]. CC BY 4.0.

Volmer–Weber (VW) or island growth: In this growth mode, the deposited material forms three-dimensional islands or clusters directly on substrate. The interaction of adsorbed atoms is much stronger among them than with substrate surface, which leads to the formation of clusters or three-dimensional islands. Thus, the deposited material does not completely wet the substrate surface and the growth is dominated by island formation.

2.3.2 Critical thickness

Critical thickness is an important parameter in heteroepitaxy, especially in lattice-mismatched heteroepitaxy, as it determines whether or not defects will form in the thin film being deposited. Critical thickness is the maximum thickness of a deposited material that can be accommodated by substrate without forming defects. Beyond this thickness, misfit dislocations start to form at the interface to relieve and relax the strain caused by lattice mismatch between substrate and deposited material. Misfit dislocations can be of different types, such as edge, screw, and mixed dislocations, depending on the lattice mismatch and crystal symmetry. Some misfit dislocations will evolve and form threading dislocations. Threading dislocations can propagate along the growth direction and reach the top epilayers, where they form a step-bunching pattern. Threading dislocations can degrade the electrical and optical properties of the layer and limit the performance of the device.

Critical thickness of a heteroepitaxy is determined by several factors, including lattice mismatch, elastic constants of the substrate and deposited material, and surface energy. Note that elastic properties could be anisotropic in cubic lattices, leading to an anisotropic lattice relaxation process and orientation-dependent thicknesses in heteroepitaxy. Critical thickness can be calculated by using various models, such as the Matthews–Blakeslee model [20, 21], which is based on the concept of strain energy relaxation. This model predicts that the critical thickness is proportional to the square of lattice mismatch, and inversely proportional to the product of elastic modulus and surface energy of the deposited material. Figure 2.3

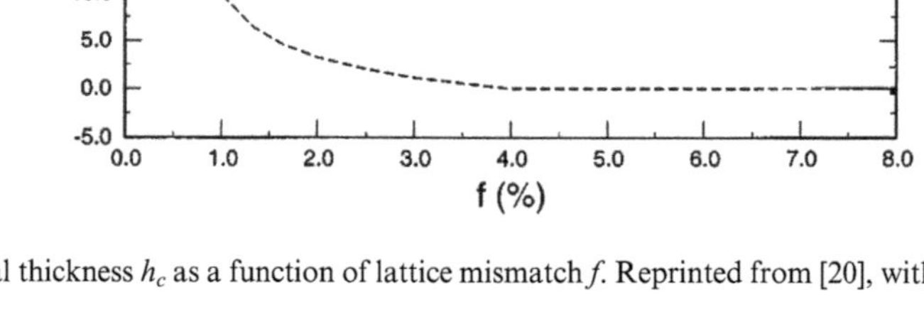

Figure 2.3. Critical thickness h_c as a function of lattice mismatch f. Reprinted from [20], with the permission of AIP Publishing.

shows the critical thickness as a function of lattice mismatch [20]. It is observed that the larger the lattice mismatch, the smaller the critical thickness. Therefore, it is challenging to achieve high-quality layer-by-layer material growth if the lattice mismatch is very large.

2.3.3 Buffer layer process

To reduce the threading dislocation density in heteroepitaxy, an intermediate buffer layer process is commonly interposed between substrates and heteroepitaxial layers to accommodate their large lattice and CTE mismatches [10–12]. A buffer process allows changes from the structure of a substrate to the structure of device layers over a few micrometer scales. The long-range change of structure allows for a gradual easing of the strain associated with the formation of an interface between two materials' lattices that don't exactly match.

Different approaches have been studied to reduce threading dislocation density efficiently:

Two-step buffer layer growth: For this approach a thin nucleation layer will initially be grown at a temperature lower than the standard growth temperature for the buffer layer in order to reduce the generation of misfit and threading dislocations. A thick buffer layer will subsequently be grown at the standard growth temperature to achieve high optical and electronic quality. This method has been reported for growing GaAs on Si with dislocation density on the order of low-10^7 cm^{-2} [22, 23].

Compositionally graded metamorphic buffer layer: For this approach, a nearly lattice-matched buffer layer is initially grown on the substrate, and then the buffer layer composition is gradually modified from the composition that is lattice matched with the substrate to a composition that is lattice matched with the top device epilayer. This compositionally graded buffer layer technology has been applied for the heteroepitaxy of III–V and SiGe semiconductors on lattice-mismatched substrates to achieve low dislocation density [24–26]. For example, a dislocation density less than 10^4 cm^{-2} has been reported for $In_xGa_{1-x}As$ ($x \leqslant 0.5$) grown on GaAs substrates [27].

Apart from reducing the generation of misfit and threading dislocations in buffer layers, another promising approach to control and reduce the ultimate dislocation density in heteroepitaxial layers is to control propagation of threading dislocations. Threading dislocations are usually formed by misfit dislocations generated near the lattice-mismatched hetero-interface. If threading dislocations propagate vertically (along the growth direction), they penetrate through buffer layer and extend into top device epitaxial layer, resulting in a high density of threading dislocations in the device active layer. However, if threading dislocations propagate laterally (parallel to the growth plane), there will be a very low density of threading dislocations within the device-active regions in top device layer. In this regard, there are three general approaches to control propagation and thus eliminate threading dislocations within the epilayers:

Strained-layer superlattice dislocation filters: In this approach, a strained-layer superlattice will be used to bend the propagation direction of threading dislocations towards the growth plane, and thus eliminate threading dislocations within the strained-layer superlattice layers before the subsequent growth of device layer, which will function as a filter for threading dislocations [22, 28].

In a recent study on the heteroepitaxy of InAs/GaAs quantum dot lasers on Si substrates, the dislocation density in the GaAs epitaxial layers was reduced from 1×10^9 cm^{-2} to the order of 10^5 cm^{-2} by using an $In_{0.18}Ga_{0.82}As$/GaAs strained-layer superlattice as a dislocation filter [22]. *In-situ* thermal annealing will also be performed to improve the efficiency of dislocation filtering by increasing the mobility of dislocations.

In-situ thermal annealing: In-situ thermal annealing can be performed during growth interruptions or after buffer layer deposition to enhance dislocation mobility [29]. By promoting glide and interaction among dislocations, annealing encourages dislocation annihilation and rearrangement, further reducing threading dislocation density and improving crystalline quality of subsequently grown layers.

Ex-situ methodologies to getter threading dislocations: Beyond growth-stage approaches, *ex-situ* dislocation gettering techniques operate by driving dislocations away from the device-active regions into non-device areas during a post-growth annealing process. This method, widely used in high-quality silicon material production, has also been adapted for the growth of HgCdTe epitaxial layers on Si substrates. For example, a rectangular mesa pattern structure was used to drive dislocations from the mesa center to the sidewalls, significantly reducing the local dislocation density [30].

As discussed above, high-quality layer-by-layer growth is essential to achieving high crystalline quality epitaxial layers with minimum defect density. Therefore, the thickness of epitaxial layers must be controlled below their critical thicknesses and various defect reduction approaches as discussed above should be considered to improve the quality of the materials.

2.4 Van der Waals heteroepitaxy

Van der Waals (vdW) epitaxy is a technique that has been revolutionizing the field of material science, particularly in the development of two-dimensional (2D) materials and heterostructures. In this type of vdW epitaxial growth, instead of traditional heteroepitaxy, a two-dimensional material, such as graphene or boron nitride, is used as a template for the growth of semiconducting material [31]. The vdW interaction occurs between two layers of materials when they are in close proximity to each other. This vdW interaction is weak, and thus the stringent requirement of lattice-constant matching for high-quality heterostructure growth in traditional semiconductor epitaxy can be relaxed since strong chemical bonding is not required at the interface. This minimizes the generation of misfit dislocations at the heterostructure interface as sketched in figures 2.4(a) and (b) [32]. This offers a unique opportunity to grow relatively high-quality thin films on 2D layered substrates without the limitation of lattice-constant matching. Another essential feature of this vdW epitaxial growth is that the layers can be easily lifted off from the 2D layered substrates and form free-standing thin films due to the vdW force

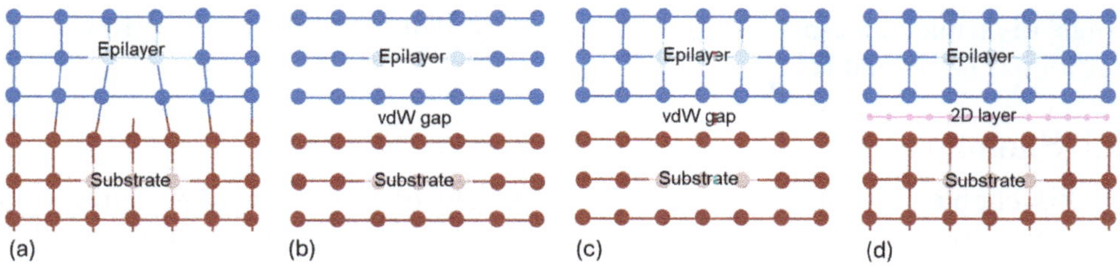

Figure 2.4. Schematics of four types of epitaxial growth models: (a) conventional heteroepitaxy of 3D material on 3D substrate, (b) vdW epitaxy of 2D material on 2D substrate, and (c) direct vdW epitaxy and (d) remote vdW epitaxy of 3D material on 2D substrate. Reprinted with permission from [32]. Copyright (2024) American Chemical Society.

between the layers. This will benefit the subsequent thin film transfer among various substrate carriers and thus allow device fabrication and testing. This potential has been recently demonstrated by researchers at IBM via the direct vdW epitaxial growth of GaN thin films on 2D graphene with low defect density for fabricating high-performance LEDs, demonstrating the potential of vdW epitaxy for growing device-quality materials [33].

Most recently, vdW epitaxy has been extended to conventional three-dimensional (3D) semiconductors. Specifically, vdW epitaxy has been investigated as a new epitaxial method for growing 3D materials on 2D layered substrates, termed as direct vdW epitaxial growth, or on 2D layer coated 3D substrates, termed as remote vdW epitaxial growth, as sketched in figures 2.4(c) and (d) [32]. For remote vdW epitaxial growth, epitaxy occurs through ultrathin 2D layers (such as graphene) between epitaxial layers and substrates since 2D layer is transparent to the Coulombic interaction between adatoms and substrate surface. Remote vdW epitaxy of single crystalline compound materials such as III–V, III–N and I–VII materials have been successfully demonstrated [13, 33–36], although remote vdW epitaxy cannot be applied for elemental semiconductors such as Si or Ge due to the lack of surface polarity. Compared with direct vdW epitaxy growth on 2D layered substrates such as Mica, remote vdW epitaxy can copy the surface orientation of the underlying 3D substrates for the subsequent 3D epitaxial layer growth. Moreover, because of the modified surface energy of the substrate surface by coating with 2D layers such as graphene, remote vdW epitaxy can promote spontaneous relaxation of heteroepitaxial films on lattice-mismatched substrates without generating defects, which is similar to what happens in the direct vdW epitaxial growth of 3D materials on 2D layered substrates. Similarly, remote epitaxial films can also be easily released from the graphene-coated substrates to form free-standing thin films, which can then be used for subsequent thin-film transfer and device fabrication and testing. Another benefit of this remote vdW epitaxial growth approach is that the graphene-coated substrates can be reused for growing more wafer batches.

Overall, lattice-mismatched heteroepitaxy based on vdW interaction is a promising approach for achieving high-quality heterostructures with low defect density and unique functional properties. The use of 2D materials, such as graphene and transition metal dichalcogenides, as substrates can enable the growth of a wide range of semiconductors and other materials that are difficult to grow using traditional heteroepitaxy.

2.5 Summary

In this chapter, we have provided an overview of the growth mechanisms of heteroepitaxy, a crucial process for the fabrication of heterostructures used in modern electronic and optoelectronic devices. We first generally introduced the concept of semiconductor heterostructures and highlighted their importance for achieving high performance semiconductor devices. We then delved into the growth mechanisms of lattice-matched heteroepitaxy and lattice-mismatched heteroepitaxy. In lattice-matched heteroepitaxy, the lattice constants of substrate and epilayer are the same, resulting in a less complicated growth process with minimal defect density. In contrast, lattice-mismatched heteroepitaxy can lead to a high density of defects, including misfit and threading dislocations. The critical thickness and three growth modes play crucial roles in the growth of epilayer, and these mechanisms must be carefully controlled to ensure a high-quality heterostructure. Finally, we discussed the growth mechanism of lattice-mismatched heteroepitaxy based on vdW interaction, which has recently gained attention due to the ability to grow 2D materials on various substrates. Looking forward, the development of new materials and growth techniques is expected to further advance the field of heteroepitaxy and enable the creation of even more complex, high quality heterostructures and thus the development of high-performance devices with new functionalities.

References

[1] Cathabard O, Teissier R, Devenson J, Moreno J C and Baranov A N 2010 Quantum cascade lasers emitting near 2.6 μm *Appl. Phys. Lett.* **96** 141110

[2] Bugajski M *et al* 2018 Mid-infrared quantum cascade lasers *Proc. of SPIE* **10974** 1097409

[3] Wang C A, Schwarz B, Siriani D F, Missaggia L J, Connors M K and Mansuripur T S 2017 MOVPE growth of LWIR AlInAs/GaInAs/InP quantum cascade lasers: impact of growth and material quality on laser performance *IEEE J. Sel. Top. Quantum Electron.* **23** 1200413

[4] Fox M and Ispasoiu R 2006 Quantum wells, superlattices, and band-gap engineering *Springer Handbook of Electronic and Photonic Materials* ed S Kasap and P Capper (Berlin: Springer) pp 1021–40

[5] Lei W, Antoszewski J and Faraone L 2015 Progress, challenges and opportunities for HgCdTe infrared materials and detectors *Appl. Phys. Rev.* **2** 041303

[6] Sato T, Mitsuhara M and Kondo Y 2009 InAs quantum-well distributed feedback lasers emitting at 2.3 μm for gas sensing applications *NTT Tech. Rev.* **7**

[7] Kim J J, Jacobs R N, Almeida L A, Jaime-Vasquez M, Nozaki C and Smith D J 2013 TEM characterization of HgCdTe/CdTe grown on GaAs (211) B substrates *J. Electron. Mater.* **42** 3142

[8] Sirtori C, Page H, Becker C and Ortiz V 2002 GaAs-AlGaAs quantum cascade lasers: physics, technology, and prospects *IEEE J. Quantum Electron.* **38** 547

[9] Troccoli M, Lyakh A, Fan J, Wang X, Maulini R, Tsekoun A G, Go R and Patel C K N 2013 Long-wave IR quantum cascade lasers for emission in the $\lambda = 8$–12 μm spectral region *Opt. Mater. Express* 1546

[10] Benson J D *et al* 2012 Growth and analysis of HgCdTe on alternate substrates *J. Electron. Mater.* **41** 2971
[11] Carmody M *et al* 2012 Recent progress in MBE growth of CdTe and HgCdTe on (211) B GaAs substrates *J. Electron. Mater.* **41** 2719
[12] Zanatta J P *et al* 2006 Molecular beam epitaxy growth of HgCdTe on Ge for third-generation infrared detectors *J. Electron. Mater.* **35** 1231
[13] Bae S H, Kum H, Kong W, Kim Y, Choi C, Lee B, Lin P, Park Y and Kim J 2019 Integration of bulk materials with two-dimensional materials for physical coupling and applications *Nat. Mater.* **18** 550–60
[14] Fukuda M 1999 *Optical Semiconductor Devices* (New York: Wiley)
[15] Razeghi M 2010 *Technology of Quantum Devices* (Boston, MA: Springer)
[16] Rogalski A, Martyniuk P and Kopytko M 2017 InAs/GaSb type-II superlattice infrared detectors: future prospect *Appl. Phys. Rev.* **4** 031304
[17] Grein C H, Boieriu P and Flatté M E 2006 Single- and two-color HgTe/CdTe superlattice based infrared detectors *Proc. of SPIE* **6127** 61270W
[18] Tassev V L 2017 Heteroepitaxy, an amazing contribution of crystal growth to the world of optics and electronics *Crystals* **7** 178
[19] https://en.wikipedia.org/wiki/Stranski%E2%80%93Krastanov_growth
[20] Dong L, Schnitker J, Smith R W and Srolovitz D J 1998 Stress relaxation and misfit dislocation nucleation in the growth of misfitting films a molecular dynamics simulation study *J. Appl. Phys.* **83** 217
[21] Braun A, Briggs K M and Böni P 2002 Analytical solution to Matthew's and Blakeslee's critical dislocation formation thickness of epitaxially grown thin films *J. Cryst. Growth* **241** 231
[22] Chen S *et al* 2016 Electrically pumped continuous-wave III–V quantum dot lasers on silicon *Nat. Photonics* **10** 307
[23] Wang Y, Wang Q, Jia Z, Li X, Deng C, Ren X, Cai S and Huang Y 2013 Three-step growth of metamorphic GaAs on Si (001) by low-pressure metal organic chemical vapor deposition *J. Vac. Sci. Technol.* B **31** 051211
[24] Lee D, Park M S, Tang Z, Luo H, Beresford R and Wieb C R 2007 Characterization of metamorphic $In_xAl_{1-x}As/GaAs$ buffer layers using reciprocal space mapping *J. Appl. Phys.* **101** 063523
[25] Qu X, Bao H, Hanieh S N, Xiong L and Zhen H 2014 An InGaAs graded buffer layer in solar cells *J. Semiconduct.* **35** 014011
[26] Currie M T, Samavedam S B, Langdo T A, Leitz C W and Fitzgerald E A 1998 Controlling threading dislocation densities in Ge on Si using graded SiGe layers and chemical-mechanical polishing *Appl. Phys. Lett.* **72** 1718
[27] Krishnamoorthy V, Lin Y W and Park R M 1992 Application of 'critical compositional difference' concept to the growth of low dislocation density ($<104/cm^2$) $In_xGa_{1-x}As$ ($x \leqslant 0.5$) on GaAs *J. Appl. Phys.* **72** 1752
[28] Shi Y *et al* 2012 Optimization of the GaAs-on-Si substrate for microelectromechanical systems (MEMS) sensor application *Materials* **5** 2917
[29] Chen Y, Farrell S, Brill G, Wijewarnasuriya P and Dhar N 2008 Dislocation reduction in CdTe/Si by molecular beam epitaxy through in-situ annealing *J. Cryst. Growth* **310** 5303
[30] Jacobs R N *et al* 2013 Analysis of mesa dislocation gettering in HgCdTe/CdTe/Si (211) by scanning transmission electron microscopy *J. Electron. Mater.* **42** 3148

[31] Bae S H, Kum H, Kong W, Kim Y, Choi C, Lee B, Lin P, Park Y and Kim J 2019 Analysis of mesa dislocation gettering in HgCdTe/CdTe/Si (211) by scanning transmission electron microscopy *Nat. Mater.* **18** 550
[32] Moon J Y, Bae S H and Lee J H 2024 Atomic spalling of a van der Waals nanomembrane *Acc. Chem. Res.* **57** 2826–35
[33] Kim J, Bayram C, Park H, Cheng C W, Dimitrakopoulos C, Ott J A, Reuter K B, Bedell S W and Sadana D K 2014 Principle of direct van der Waals epitaxy of single-crystalline films on epitaxial graphene *Nat. Commun.* **5** 4836
[34] Kim Y *et al* 2017 Remote epitaxy through graphene enables two-dimensional material-based layer transfer *Nature* **544** 340–3
[35] Kong W *et al* 2018 Polarity governs atomic interaction through two-dimensional materials *Nat. Mater.* **17** 999
[36] Badokas K, Kadys A, Mickevicius J, Ignatjev I, Skapas M, Stanionyte S, Radiunas E, Juska G and Malinauskas T 2021 Remote epitaxy of GaN via graphene on GaN/sapphire templates *J. Phys.* D **54** 205103

Chapter 3

Heteroepitaxial growth of HgCdTe on lattice-mismatched substrates

In chapter 2, we discussed the growth mechanism of general heteroepitaxial growth including lattice-mismatched heteroepitaxial growth. The main challenge for lattice-mismatched heteroepitaxial growth is the lattice mismatch which can generate misfit dislocations around the interface between substrate and epilayer. These misfit dislocations can form threading dislocations which can penetrate into the top epilayers and thus deteriorate the performance of the device made out of the top epilayer. Thus, it is essential to invent and study various dislocation reduction techniques to achieve high-quality epilayers on lattice-mismatched substrates. This chapter will focus on the heteroepitaxial growth of high-quality CdTe and HgCdTe layers on lattice-mismatched alternative substrates including Si, Ge, GaAs, and GaSb. Special attention will be given to dislocation reduction techniques with the goal of achieving high-quality HgCdTe epilayers with low threading dislocation density that are suitable for making high-performance LWIR HgCdTe detectors.

3.1 Introduction to heteroepitaxial growth of HgCdTe on lattice-mismatched substrates

As introduced in chapter 1, since their origin in 1959 [1], HgCdTe IR detectors have undergone fast development from the initial photoconductor structure with a scanning imaging function, to the current state-of-the-art photodiode structure with a starring imaging function. These have been widely applied in various industry sectors due to superior device and imaging performance, including night vision,

doi:10.1088/978-0-7503-3443-3ch3

© IOP Publishing Ltd 2025. All rights, including for text and data mining (TDM), artificial intelligence (AI) training, and similar technologies, are reserved.

target identification, metrology, earth remote sensing, space astronomy study, and many others. Despite the huge advances made in HgCdTe IR detector technology, the current state-of-art HgCdTe IR detectors and their FPAs on the market suffer some serious limitations such as higher cost and smaller array format size, which substantially limit their broad industry applications. For example, the current mainstream array format size for MWIR HgCdTe FPAs is 640 × 512, and the unit cost is very high (~$35 000 per unit). Although HgCdTe FPAs with a larger array format size were reported by some large defence/aerospace corporations (4096 × 4096 is the largest format size reported so far [2]), it is very challenging to fabricate them (only Teledyne reported the relevant product), and the cost is extremely high (~million dollars per unit for FPAs with a million pixels). Obviously, the cost of HgCdTe IR detectors is much higher while the array format is much smaller in comparison to those of the current state-of-the-art complementary metal-oxide semiconductor (CMOS) imagers available on the market.

As discussed in chapter 1, the further development of IR applications requires the future IR detectors and their FPAs to have a larger array format and relatively lower cost, which present as two features of the next generation or 'third generation' IR detectors, and have been the long-term goal of the IR community [3–5]. Once again, as discussed in chapter 1, the main single dominant factor that impacts on the higher cost and smaller array format size of HgCdTe-based FPAs is the CdZnTe substrates used to grow these IR detectors. The CdZnTe substrates suffer the main limitations of lower crystal quality, smaller wafer size, higher cost, and limited commercial availability. Therefore, there is a strong impetus to replace CdZnTe substrates with alternative high-quality, large-area, and low-cost substrates in order to enhance FPA yield, increase wafer size and, thus increase the number of fully functional FPAs per wafer in order to reduce the overall cost.

Over the past two decades, several alternative substrates have been studied for replacing CdZnTe for growing HgCdTe materials, such as Si, Ge, GaAs, and GaSb [6, 7]. All these substrates have the features of high crystal quality (in terms of EPD numbers), large wafer size, lower cost, and ready availability. However, all these alternative substrates present a very large lattice and CTE mismatch with HgCdTe materials. As shown in figure 1.7, Si, Ge, GaAs, and GaSb substrates present a lattice mismatch of 19%, 14.3%, 14.4%, and 6.1% to $Hg_{0.7}Cd_{0.3}Te$, and a CTE mismatch of 80%, 20%, 14%, and 23% to $Hg_{0.7}Cd_{0.3}Te$. As discussed in chapter 2, this lattice mismatch between HgCdTe and substrates will result in the accumulation of misfit strain (elastic) with increasing thickness of the top HgCdTe layer during epitaxial growth. Once the top HgCdTe layer exceeds its critical thickness, the accumulated strain will be relaxed to stabilize the system by forming misfit dislocations ($>10^8$ cm^{-2}) in the vicinity of the interface [8]. These misfit dislocations can form threading dislocations which propagate vertically (along the growth direction) to form a high density of threading dislocations in the top HgCdTe epilayer if grown directly on these alternative substrates [9]. These dislocations will result in defect energy levels, as well as carrier scattering and recombination centers in the HgCdTe epitaxial layers. This will not only reduce the carrier mobility and minority carrier lifetime (and thus degrade detector performance), but it also leads to a high density of defective/dead pixels in the FPAs, which lowers the FPA yield and increases the production cost.

To reduce/minimize the dislocation density in the HgCdTe epilayers, a CdTe buffer layer is generally introduced between the HgCdTe and the lattice-mismatched alternative substrates to relax and block the penetration of dislocations propagating from the lattice-mismatched substrate/CdTe interface, the sample structure of which is shown in figure 3.1(a) [10]. From figure 3.1(a), CdTe has a lattice constant almost the same as HgCdTe, and thus there will also be a large lattice and CTE mismatch between CdTe and those alternative substrates. Similarly, the lattice-mismatched strain will lead to the formation of a large number of misfit dislocations around the interface between CdTe and alternative substrates, which will also form threading dislocations which can propagate along the growth direction during the epitaxy. However, CdTe buffer layers typically grow very thick (>5 μm). As a result, some threading dislocations will interact with other dislocations, merge, and be annihilated while they propagate to the surface, which will significantly reduce the number of threading dislocations reaching the top HgCdTe epilayers, and thus effectively reduce the dislocation density in the top HgCdTe epilayers. Figure 3.1(b) shows the relationship between the full width at half maximum (FWHM) of the x-ray diffraction (XRD) peak of the CdTe buffer layer as a function of the CdTe buffer layer thickness when grown on Si and GaAs [11]. It can be observed that a CdTe buffer layer with an appropriate thickness (>10 μm) can effectively block the penetration of dislocations, leading to high crystal quality of the CdTe buffer layer grown on GaAs (XRD FWHM ~60 arcsec). In the meantime, it is also observed that the EPD also decreases with increasing the CdTe buffer layer thickness, and the EPD saturates around low-10^6 cm^{-2} to mid-10^6 cm^{-2} when the CdTe buffer layer thickness is over 10 μm. It should be noted that using a thick CdTe buffer layer is the

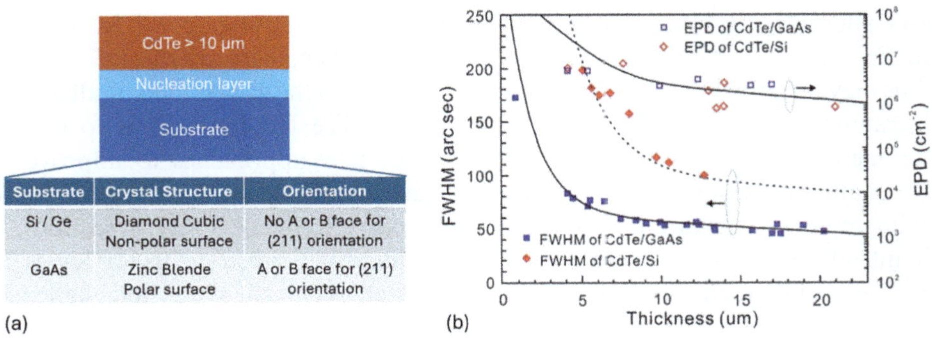

Figure 3.1. (a) General schematic sample structure for growing HgCdTe/CdTe on alternative substrates. Reproduced from [10], with permission from Springer Nature. (b) XRD FWHM and EPD values for CdTe buffer layers grown on GaAs and Si in relation to CdTe buffer layer thickness. Reprinted from [11], Copyright © 2006, with permission from Elsevier B.V. All rights reserved.

most common approach for reducing the dislocation density in the subsequent HgCdTe epilayers and generally leads to an EPD level of mid-10^6 to low-10^7 cm^{-2} which is the current status for Si, Ge, and GaAs alternative substrate technologies [11–13]. Essentially, a CdTe buffer layer provides the structural template necessary for the subsequent HgCdTe growth, and dislocations that originate within the CdTe buffer serve as the baseline for the minimum dislocation density that will be present in the overlying HgCdTe layer. Therefore, it is essential to have CdTe buffer layer with minimum dislocation density.

Although the dislocation density in HgCdTe epilayers grown on Si, Ge, and GaAs has been reduced to the level of mid-10^6 to low-10^7 cm^{-2} after two decades of effort [11–13], it is still almost two orders of magnitude higher than that for the counterpart material grown on lattice-matched CdZnTe substrates [12]. Although such high dislocation densities can be tolerated by SWIR and MWIR HgCdTe detectors, they significantly degrade the performance of LWIR HgCdTe detectors because LWIR devices are more sensitive to the defects due to their narrower bandgap. For dislocation densities $<5 \times 10^5$ cm^{-2} the dislocation density has little effect on the minority-carrier recombination time, even for LWIR HgCdTe, because carrier recombination is dominated by intrinsic Auger mechanism. However, for dislocation densities above the mid-10^5 cm^{-2} range, carrier recombination is dominated by Shockley–Read–Hall (SRH) recombination due to the dislocations, and the minority-carrier recombination time for LWIR HgCdTe is roughly inversely proportional to the dislocation density [14]. Also, dislocations can locally short out p–n junctions in devices, considerably increasing the dark current [15]. As a result, the MBE growth of HgCdTe on lattice-mismatched alternative substrates yields SWIR and MWIR photodiodes with performance similar to those grown on CdZnTe substrates at operating temperatures above 77 K [16–19]. However, their LWIR or VLWIR devices present significantly reduced performance in comparison to the counterpart devices grown on CdZnTe substrates [17, 20]. Generally, to achieve high performance in LWIR HgCdTe FPAs, the dislocation density in HgCdTe needs to be less than 5×10^5 cm^{-2} [21]. This renders current Si, Ge, and GaAs alternative substrate technologies unsuitable for growing LWIR HgCdTe epitaxial materials, and is currently the main challenge to alternative substrate technology for growing HgCdTe. Therefore, the key to the application of alternative substrates for fabricating high performance HgCdTe IR detectors is to develop either new alternative substrate technologies or new approaches to improve the existing alternative substrate technologies to meet the dislocation density criteria for making high-performance LWIR HgCdTe detectors. In the next several sections, we will introduce the development of various alternative substrate technologies for HgCdTe, including the development history as well as current status and challenges.

3.2 Heteroepitaxial growth of CdTe and HgCdTe on Si substrates

Over the past two decades, various group IV and III–V substrates have been proposed and studied as alternative substrates for growing HgCdTe such as Si, Ge,

GaAs, and GaSb [11–13]. Among them, Si as an alternative substrate has attracted more attention due to its compatibility with the Si readout circuit (ROIC) in a flip-chip bonded configuration. This provides the possibility of fabricating monolithic detectors and FPAs in which the functions of detection and charge readout would be monolithically integrated onto a single Si wafer in the ultimate FPA. This possibility has been demonstrated by the epitaxial growth of HgCdTe in selected areas of a Si ROIC [22]. Apart from potential monolithic integration, Si alternative substrates also provide other advantages/benefits in comparison to other alternative substrates, including [6]: (1) Si wafers are commercial wafers with much larger sizes (maximum available size of 11.8 inches in diameter). This allows either FPAs with much larger format size (greater imaging resolution) or greater numbers of smaller FPAs for reduced production costs. (2) Si substrate has perfect thermal matching of its lattice to that of the ROIC, and thus no thermal mismatch limit on the size of FPAs. Note that thermal mismatch between substrate and Si ROIC generates a strain upon cycling down to cryogenic operating temperatures which could destroy the Indium bump bonding for very large-area FPAs. This could be a challenge for other alternative substrates. (3) Si substrates allow the use of automated Si processing technology for increased efficiency in device fabrication and reduced production costs. (4) Si substrate is the most robust and durable substrate material, leading to less breakage and a higher device yield, and thus reduced production costs. (5) Si has the highest thermal conductivity, leading to the highest lateral uniformity for MBE-grown HgCdTe in terms of both composition and thickness. (6) Si offers the lowest level of impurity migration into the HgCdTe materials. (7) Si substrates have a very high crystal quality in terms of having the lowest density of surface defects. However, it should be noted that Si has the worst lattice and thermal mismatches with HgCdTe in comparison to any other substrate materials considered for growing HgCdTe. As shown in figure 1.7, Si substrates have a lattice constant mismatch of 19% with HgCdTe which is much larger than those of Ge (~14.3%), GaAs (~14.4%) and GaSb (~6.1%), and forms the major challenge confronted by the growth of CdTe and HgCdTe layers on Si.

The MBE growth of CdTe and HgCdTe on Si has been studied over more than two decades mainly in the United States and China with the first study reported in 1989 [23]. The primary challenge to the MBE growth of Te-based II–VI compounds on Si (211) is the nucleation of the II–VI layer. For growing CdTe and HgCdTe on Si substrates, a ZnTe nucleation layer and an As passivation layer are commonly applied to suppress the formation of micro-twin defects and to improve the crystal quality of the CdTe buffer layer and subsequently grown HgCdTe epilayer. A typical growth procedure is as follows [24]: Si wafers are cleaned using an RCA cleaning process (or modified RCA cleaning process) which leaves a very thin uniform oxide layer on the Si surface. Then, the samples are quickly heated up to a proper temperature (e.g., 1050 °C) to remove the oxide layer and then quickly cooled under an As flux to 500 °C. Finally, the sample is cooled to the nucleation temperature of 340 °C without any flux. After that, a nucleation layer of thin ZnTe is deposited at 340 °C on Si substrate utilizing migration-enhanced epitaxy

(MEE) [25] with elemental Zn and Te sources. Then, the ZnTe layer is annealed at 490 °C with Te flux for 20 min to enhance the nucleation process. After the annealing, a thick (>5 μm) CdTe layer is grown at a proper growth temperature (typically between 280 °C and 320 °C) with a proper growth rate (usually ~1 μm h^{-1}) using CdTe and Te sources or Cd and Te sources.

After the growth of CdTe buffer layer, the samples can be directly transferred to the HgCdTe growth chamber for growing HgCdTe without breaking the vacuum if using a dual-chamber MBE system. If a dual-chamber MBE system is not used (in most cases), the samples (Si substrates with CdTe buffer layers on the top) must be surface prepared for HgCdTe growth using the same surface preparation process as that of CdZnTe substrates. The surface preparation process can be briefly described as follows [6]: the samples are first degreased in two separately heated trichloroethylene baths followed by rinses in two methanol baths. They are then prepared for HgCdTe growth by etching in a bromine/methanol solution, followed by several methanol and deionized water rinses. This leaves a Te-rich surface. Finally, the samples are dried with nitrogen gas and loaded into the MBE system. After that, the samples will go through the thermal cleaning process (thermal desorption of oxide on the surface), and then cool down to the proper temperature (e.g., 185 °C) for growing HgCdTe. The thermal cleaning process is usually monitored by reflection high-energy electron diffraction (RHEED). (Typically, a streaky RHEED pattern with rods indicates successful thermal cleaning of the CdTe surface.) After achieving high-quality thermal cleaning of CdTe/Si substrates, HgCdTe growth on CdTe/Si substrates can be undertaken in a way similar to that on CdZnTe substrates: a Hg flux is first present even before the growth shutters are opened, so that the growth starts with a monolayer of Hg. The growth proceeds at a proper growth rate (usually 2–3 μm h^{-1}) with appropriate fluxes of Hg, CdTe, and Te (e.g., a flux of 3×10^{-4} Torr for Hg, 10^{-6} Torr for Te, and 2×10^{-7} Torr for CdTe). Note that at some institutions, an additional thin CdTe layer is grown on CdTe/Si substrates and annealed before the growth of HgCdTe. This procedure ensures a smooth and clean initial surface prior to HgCdTe nucleation.

During the growth of HgCdTe, RHEED can also be used for monitoring the growth process. Note RHEED is a common tool installed on most MBE systems, while some MBE systems use spectral ellipsometer (SE) for monitoring the growth. Figure 3.2 shows the typical schematic RHEED patterns and the corresponding features on the growth front, as well as possible changes of the growth conditions to be undertaken [26]. For MBE systems with *in-situ* SE tools, the monitoring process is made easier by using the data library provided by the SE manufacturers: the growth temperature can be continuously adjusted to minimize the roughness of the growing surface as measured by SE, and the CdTe flux can be continuously adjusted to maintain the SE-measured composition at the targeted value. To obtain lateral uniformity in the growth, the sample is continuously rotated as the growth fluxes are incident off center at an angle to the surface normal.

With the above special growth procedures involving As-passivation and ZnTe nucleation layers, as-grown CdTe and HgCdTe layers typically show an EPD

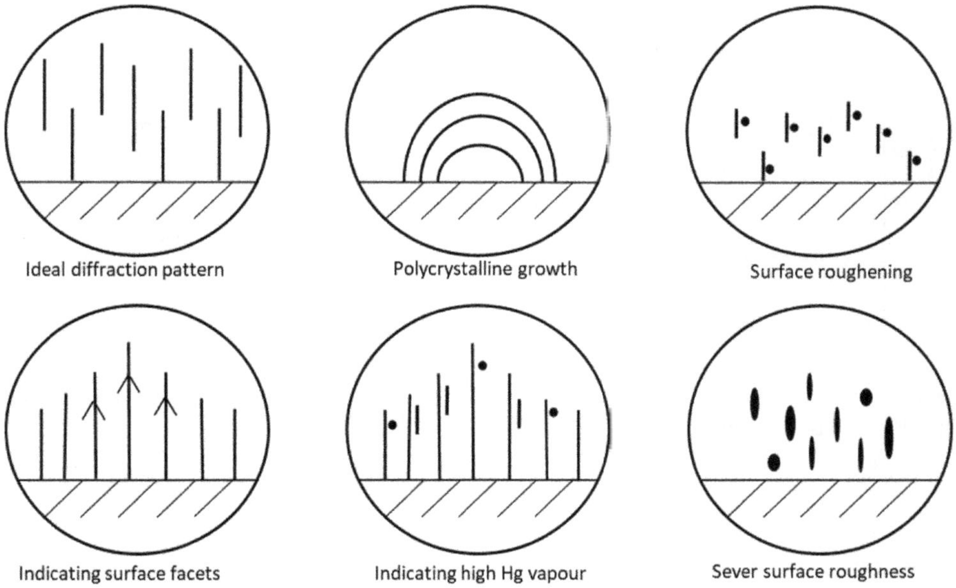

Figure 3.2. Schematic RHEED diffraction patterns for MBE growth of HgCdTe. (Reproduced from [26] under the permission of UWA open access policy.)

around the level of mid-10^6 cm^{-2} to low-10^7 cm^{-2} which does not meet the EDP criteria required for making high-performance HgCdTe IR detectors, especially LWIR IR detectors. As a result, determining how to reduce the dislocation density and/or lower the impact of dislocations on the device performance become critical for their ultimate applications in high-performance IR detectors. In principle, there are three general approaches to reduce the dislocation density in CdTe and HgCdTe layers and/or lower the impact of dislocations on their device performance: (1) reduction of as-grown dislocation density; (2) reduction of dislocation density via post-growth treatment; and (3) passivation of the defects to reduce their impact on device performance.

3.2.1 Approaches to reduce the as-grown dislocation density

There are a number of effective approaches which can be used to reduce the as-grown dislocation density in CdTe buffer layers and HgCdTe materials. These include:

In-situ cycling annealing: Recent studies on MBE growth of CdTe/Si showed that *in-situ* thermal cycle annealing during the MBE growth process itself is an effective approach to reduce EPD and improve overall crystal quality of the CdTe layers grown [27]. Figure 3.3 shows the schematic annealing temperature profile, the detailed annealing conditions of which can be found in reference [27]. Figure 3.4(a) shows the XRD FWHM of the CdTe/Si layers as a function of the number of

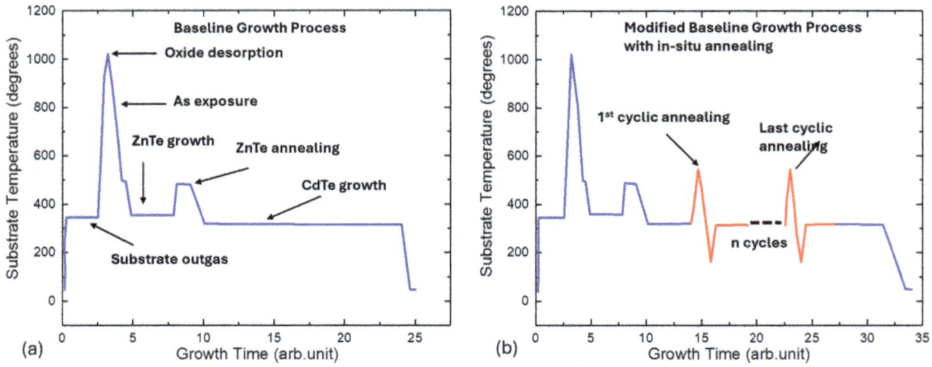

Figure 3.3. Temperature profiles for an MBE-growth process of CdTe on Si with *in-situ* cyclic annealing. Reprinted from [27] Copyright © 2006, Published by Elsevier B.V.

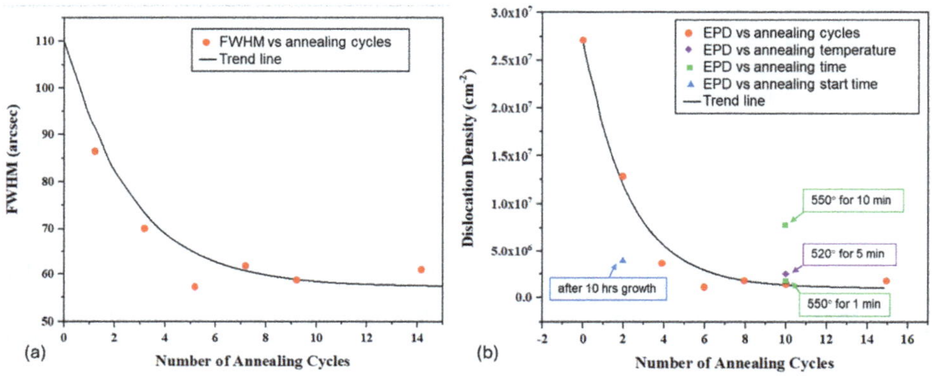

Figure 3.4. (a) XRD FWHMs of CdTe layers as a function of the number of *in-situ* annealing cycles, (b) Dislocation densities of CdTe/Si layers grown with different *in-situ* annealing cycles and other annealing conditions. Solid line is exponential trend line fit to the data of EPD vs. number of annealing cycles. Reprinted from [27] Copyright © 2006, Published by Elsevier B.V.

annealing cycles. It is observed that the XRD FWHM monotonically reduces with increasing the number of annealing cycles from 0 to 6, but remains unchanged as the number of cycles further increases. This indicates that adding cyclic annealing during the growth does improve the overall crystalline quality of CdTe/Si layers. Figure 3.4(b) shows the EPD of the same set of samples as a function of the number of annealing cycles, labeled by solid circles. It is observed that the dislocation density of CdTe/Si reduces exponentially as the number of annealing cycles increases. A reduction of two orders of magnitude is achieved for the dislocation density (EDP of 2.7×10^7 cm^{-2} for CdTe/Si without cycle thermal annealing and 4.3×10^5 cm^{-2} for that with 10 cycles' thermal annealing). This clearly implies that *in-situ* cyclic thermal annealing is very effective in suppressing the threading dislocations in CdTe/Si.

Apart from the number of annealing cycles, thermal annealing conditions also present influence on the dislocation reduction results. Figure 3.4(b) shows the influence of annealing duration on the dislocation reduction results (represented by the solid squares in figure 3.4(b)). It is observed that a 5-min annealing duration at 550 °C for 10 cycles shows the most effective dislocation reduction in comparison to an annealing duration of both 1 and 10 min. It is understood that thermal annealing provides a mechanism for those as-grown dislocations to move caused by the thermal energy and the strain field formation due to the rapid change of layer temperature. The duration of 1 min is too short, and the dislocations do not have sufficient time to interact and annihilate, leaving a higher density of threading dislocations. However, the duration of 10 min is too long, which can cause damage to the CdTe/Si due to too much thermal energy added. Therefore, a 5 min duration gives the best result in this experiment.

Similarly, the annealing temperature shows an impact on the dislocation reduction results as indicated by the solid diamonds in figure 3.4(b). It is observed that higher annealing temperature produces better dislocation reduction results: EPD of 4.3×10^5 cm^{-2} for annealing at 550 °C and 2.0×10^6 cm^{-2} for annealing at 520 °C with the same number of annealing cycles. In addition, the experiments also indicate that thermal annealing initiated at a later growth stage has a better chance to reduce threading dislocations, which is labeled by the solid triangle in figure 3.4 (b). For the same two cycles of thermal annealing, the layer having annealing started 10 h after CdTe growth shows a much lower dislocation density than the layer with annealing started after only 5 h of CdTe nucleation. If most of the dislocations initiate at the layer/substrate interface, the dislocations threaded to the top portion of the thicker layer have longer dislocation lines, and thus will have a stronger tendency to move under favorable conditions, and a better chance to encounter a free surface or another dislocation, leading to more effective dislocation reduction.

To better understand the mechanism of dislocation reduction in the CdTe/Si system with thermal cyclic annealing, EPD etching experiments were undertaken to study the dislocation density as a function of layer thickness. Figure 3.5 shows the dislocation densities of CdTe/Si layers grown with and without *in-situ* cyclic annealing as a function of layer thickness. With 10 cycles of annealing, continuous annihilation and fusion of the dislocations is observed beyond 3 μm and then dislocation density reaches a saturation level ($\sim 5 \times 10^5$ cm^{-2}) at about 5 μm from the interface. Note that a dislocation density of 5×10^5 cm^{-2} indicates that the average separation distance among the dislocations is 22 μm, which is an order of magnitude larger than the annihilation and fusing radius (1.8 μm) for CdTe/Si based on Speck's model [28]. This shows that the thermal energy and thermal strain generated during temperature cycle annealing can mobilize threading dislocations and reduce the separation of the dislocation below the critical radius, leading to annihilation or fusion of two dislocations. This is supported by the fact the number of dislocations decreases with both an increase in the number of annealing cycles and an increase in annealing temperature as shown in figure 3.4(b).

It has been observed that subjecting CdTe/Si layers to multiple-cycle *in-situ* annealing can result in a reduction of EPD down to a saturation point of mid-10^5 cm^{-2}. It should be noted that this *in-situ* annealing process is generally limited to the growth of CdTe layer, not HgCdTe. This is because high temperature annealing can cause a rough surface and even degrade the surface if done without appropriate

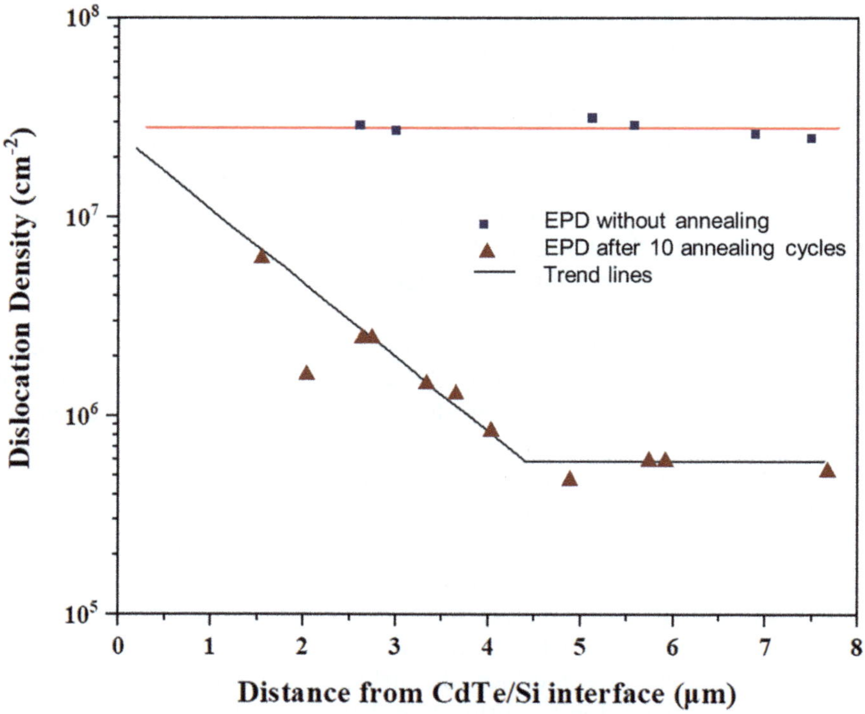

Figure 3.5. Dislocation densities of CdTe/Si layers grown with and without *in-situ* cyclic annealing as a function of layer thickness. Solid lines are trendlines. Reprinted from [27] Copyright © 2006, Published by Elsevier B.V.

overpressure protection. However, for *in-situ* annealing configuration in the MBE chamber, the Hg and Te fluxes are very limited, which might not be appropriate for maintaining a high-quality sample surface without damage. Therefore, such *in-situ* annealing was not typically used for improving crystal quality and reducing the dislocation density of HgCdTe epilayers. Even for CdTe buffer layers, *in-situ* annealing is also limited by the protective fluxes and other annealing conditions, and thus is not widely applied like the *ex-situ* annealing process.

Strained-layer superlattice pre-buffer layer: A ZnTe/CdTe strained superlattice pre-buffer layer could be introduced before the growth of the CdTe layer with the aim to relax the lattice mismatch strain and thus reduce defect generation around the interface between the Si substrate and the CdTe layer [6]. This will result in a lower dislocation density in the CdTe buffer layer and the subsequent HgCdTe layer. Despite this potential, there are no reports of its occurrence due to the challenge in growing high-quality ZnTe/CdTe strained superlattices.

Lattice-matched buffer layers: The CdTe buffer is the simplest and most widely used buffer layer for growing HgCdTe on alternative substrates; however, it is slightly lattice-mismatched to HgCdTe. In this regard, some buffer layers with lattice constants matched to HgCdTe were proposed and studied with the aim of reducing the defect density in the subsequent HgCdTe layers, including

$Cd_{0.96}Zn_{0.04}Te$ [6, 29], CdSeTe [6, 30], CdZnSeTe [6, 31], and Be chalcogenides (BeCdTe or BeMgTe $BeSe_{0.45}Te_{0.55}$) [6]. Although these buffer layers offer the advantage of an almost perfect lattice match to the HgCdTe to be grown, they were not widely applied for industry processing because of the difficulty of obtaining epilayers of sufficient crystalline quality and/or the potential contaminant to the MBE growth chamber.

Improved surface preparation methods: Some improved surface preparation methods were studied to achieve better surface thermal cleaning results and reduce the defect density in the CdTe and HgCdTe layers. These improved methods include: (1) *surface re-oxidation* [6]. For MBE growth of CdTe on Si, (211) crystal orientation is typically used due to a higher Hg sticking coefficient. However, Si wafers with (211) crystal orientation present a high density of surface steps, which makes it hard to remove oxygen and other contaminants from the surface. The purpose of surface re-oxidation is to oxidize surface contaminants to facilitate their low-temperature removal, which will benefit polarity control of the II–VI layers grown subsequently, and lead to dislocation density reduction within. As reported in reference [6], Si surface can be first cleaned in a mixture of NH_4OH:H_2O_2:H_2O (1:1:5, 85 °C), etched in dilute HF (2%) to remove the oxide, and then re-oxidized in a solution of HCl:H_2O_2:H_2O (1:1:5 at 65 °C); (2) *hydrogen passivation* [6]. Hydrogen passivation was also studied for improving the material quality of CdTe layers grown on Si. A hydrogen passivation layer can be volatile around 560 °C, and thus greatly reduces the oxygen and carbon contamination of the Si surface without an extensive prebaking or high-temperature thermal desorption usually required after RCA cleaning. These improved surface preparation methods allow the growth of epitaxial layers with reduced dislocation densities, but complicate the surface preparation process. As a result, they are not widely applied.

So far, the above methods have yielded qualitative, but not quantitative improvements in the material quality of CdTe and HgCdTe epilayers grown on Si substrates. The as-grown EDPs in CdTe and HgCdTe layers are still not low enough for making high-performance HgCdTe IR detectors. In the following sections, we will discuss post-growth treatments to either reduce the defect density or passivate the defects in the CdTe and HgCdTe epilayers.

3.2.2 Approach to reduce dislocation density via post-growth treatment

Apart from high-temperature *in-situ* cyclic annealing, high-temperature *ex-situ* cyclic annealing was also studied and demonstrated to be an efficient approach to reduce the dislocation density in CdTe and HgCdTe epilayers. In one reported study [32], the as-grown HgCdTe/CdTe/Si samples were sealed in a closed ampoule together with Hg droplets, and then underwent high-temperature cyclic annealing processes. Note that the closed ampoule with Hg droplets inside was to make sure the annealing was undertaken in a mercury atmosphere. To understand the defect reduction mechanism and achieve the best defect reduction results, various annealing parameters were studied including Hg overpressure, annealing temperature set-point, number of annealing cycles, and total annealing time. Experimental samples were also characterized using x-ray diffraction and Hall measurement to understand the impact of *ex-situ* cyclic annealing on the crystalline quality and electrical properties of HgCdTe materials, which will impact the ultimate detector performance. Figure 3.6 shows a schematic cycle annealing experiment process reported [32].

It was found that proper Hg overpressure was required to maintain a smooth and damage-free sample surface for annealing, especially at higher annealing temperatures. Typically, a significantly higher Hg overpressure is required for a higher annealing temperature. For example, it was found that triple the amount of Hg was required for 450 °C/four cycle anneals with respect to the counterpart 400 °C anneals. It was easier to either roughen the surface (too little Hg) or damage the surface through condensation defects (too much Hg). Figure 3.7 shows the one-to-one surface images of HgCdTe/Si before and after cycle annealing [32]. It can be observed that a smooth and clean sample surface can be maintained after high-temperature cyclic annealing by using proper Hg overpressure (e.g., the proper amount of Hg in the sealed ampoule).

Figure 3.8(a) shows the dislocation density with respect to the number of annealing cycles (400 °C cycle anneal with the baseline Hg amount) [32]. It is found that the dislocation reduction is directly related to the number of annealing cycles rather than the total duration of annealing time, and that the decrease in dislocation density follows an exponential decay as the number of cycles increases. The dislocation density reaches a minimum value somewhere between four and eight

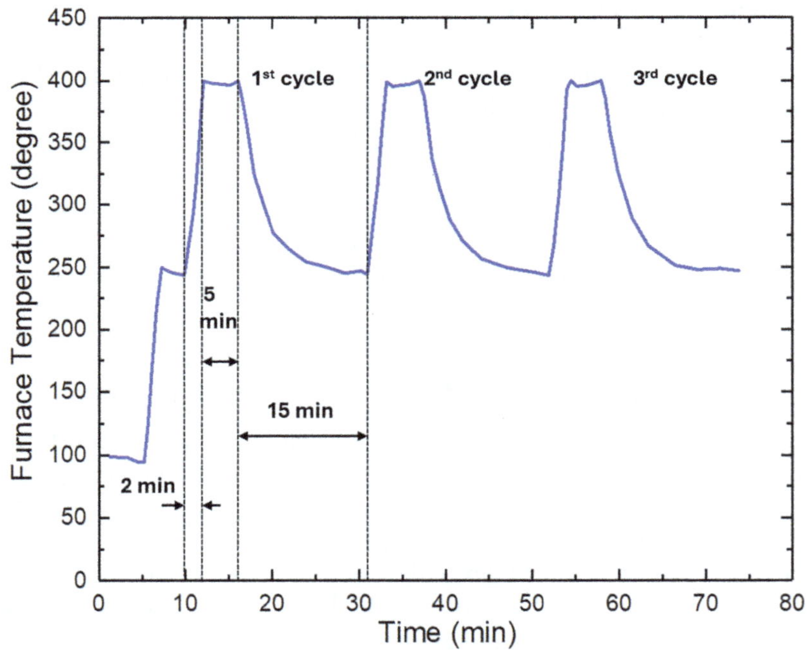

Figure 3.6. Temperature profile of a three-cycle *ex-situ* annealing experiment conducted with a maximum furnace temperature setpoint of 400 °C. Annealing temperature, annealing duration, and number of cycles were variables within this study. Reproduced from [32] with permission from Springer Nature.

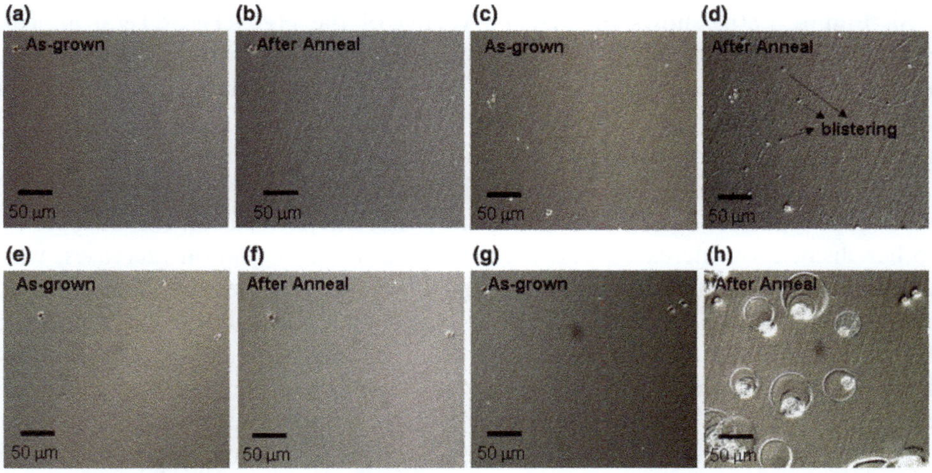

Figure 3.7. One-to-one surface images of HgCdTe/Si before and after cycle annealing: (a, b) 400 °C/four-cycle anneal using the baseline Hg amount; (c, d) 450 °C/four-cycle anneal using the baseline Hg amount; (e, f) 450 °C/four-cycle anneal using three times the baseline Hg amount; and (g, h) 450 °C/four-cycle anneal using four times the baseline Hg amount. Reproduced from [32] with permission from Springer Nature.

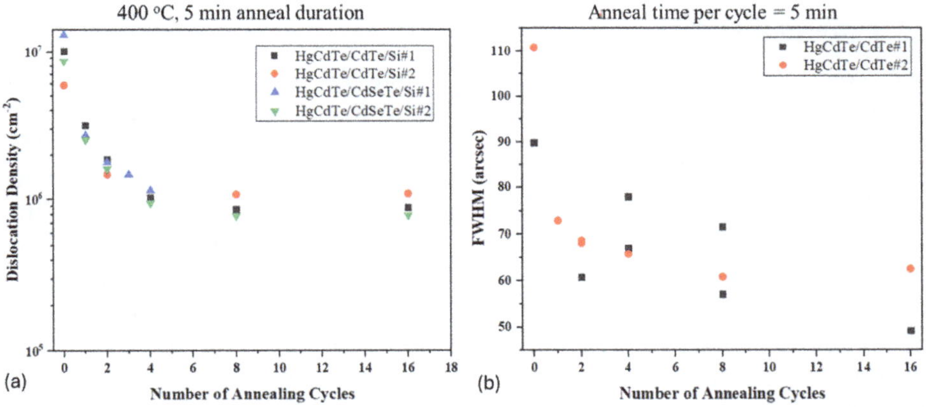

Figure 3.8. (a) Dislocation reduction as a function of the number of annealing cycles. Four layers were studied: two grown on CdTe/Si composite substrates and two grown on lattice-matched CdSeTe/Si composite substrates. The data points at zero cycles correspond to the as-grown dislocation density, (b) Variation of XRD FWHM of the (422) crystal plane with number of annealing cycles. Reproduced from [32, 33] with permission from Springer Nature.

cycles, and no further reduction is obtained with even more cycles added. As shown in figure 3.8(a), the dislocation reduction is observed to saturate at a density of $\sim 1 \times 10^6$ cm^{-2}. From these results, it can be seen that the main mechanism for dislocation reduction is the relative change in temperature that the sample experiences during the cycle annealing process.

Apart from the reduction in dislocation density, high-temperature cyclic anneals also lead to an improvement in crystalline quality of the HgCdTe materials grown on Si. Figure 3.8(b) shows the XRD FWHM of the HgCdTe/CdTe/Si annealed samples in comparison with the as-grown layer [33]. It can be observed that annealing causes the reduction in XRD FWHM, indicating an improved crystalline quality of the material. The main mechanism for dislocation reduction upon cyclic anneals is that the temperature change (both positive and negative) induces a strain field within the material, which in turn leads to dislocation movement. Once the dislocations are able to move, the probability that one dislocation can interact with another dislocation increases, and hence dislocation annihilation increases, which will enhance the crystalline quality of HgCdTe/CdTe/Si. However, after a certain point, the dislocations become isolated far enough that one dislocation can no longer interact with another. At this point, the overall material dislocation value reaches a minimum and saturates even as more annealing cycles are added, which also indicates the material's crystalline quality will not improve further.

As discussed above, with *ex-situ* cyclic annealing dislocations have enough energy to move around, and then merge and finally annihilate, leading to the significant reduction of dislocation in the materials. The problem with this approach is that high-temperature annealing can cause substantial diffusion and destroy the sharpness of the hetero- and homo-junctions required by many modern detector designs.

3.2.3 Passivation of dislocations

Apart from reducing defect density in CdTe and HgCdTe layers, another effective approach to reduce the impact of defects on the device performance is to passivate the defects, i.e., passivate and reduce the trap states caused by defects [6, 34, 35]. In this regard, hydrogen passivation was studied for passivating the dislocations in HgCdTe materials grown on lattice-mismatched alternative substates [34]. In the study reported by Boieriu *et al* [34], LWIR HgCdTe materials grown on Si substrates were exposed to hydrogen plasma generated by electron cyclotron resonance (ECR). Hall measurements and minority carrier lifetime measurements were undertaken to study the impact of hydrogen passivation on the major electrical properties of HgCdTe—carrier mobility and minority carrier lifetime which determine detector performance such as quantum efficiency. HgCdTe/Si samples with different EPD levels were studied to better understand the impact of hydrogen passivation on electrical properties. It was found that no matter what the EPD level was, the carrier mobility and minority carrier lifetime of the LWIR HgCdTe samples all improved, and the HgCdTe/Si sample with the highest EPD level presented the largest change in conduction property. As dislocations in the HgCdTe materials could act as trap states and scattering centers, and thus degrade the carrier mobility and minority carrier lifetime, the improvement in the carrier mobility and minority carrier lifetime of the LWIR HgCdTe samples suggested the dislocations and thus trap states in HgCdTe materials were successfully passivated after exposure to the ECR hydrogen plasma. Note that the minority carrier lifetime of HgCdTe was improved by one order of magnitude after exposure to the hydrogen plasma, indicating an efficient passivation. Secondary ion mass spectrometry (SIMS) profile measurements of hydrogen of these HgCdTe/Si samples also confirmed the incorporation of hydrogen in these samples. This efficient defect passivation was mainly due to the efficient bonding of hydrogen atoms with dangling bonds formed along the dislocations and thus the elimination of trap states. This led to the improvement in the carrier mobility and minority carrier lifetime as observed. This hydrogen passivation was quite stable as there was no change in the carrier mobility and minority carrier lifetime after 3 months' shelf storage and after being heated and stored at 80 °C for 40 min. Similar hydrogen passivation effect of defects was also observed in LWIR HgCdTe materials grown on GaAs substrates by a research group from SITP [35]. All these suggest that hydrogen passivation provides another effective approach to reduce the impact of defects in CdTe and HgCdTe materials grown on lattice-mismatched alternative substrates.

Apart from the above defect reduction/passivation methods, some other methods were also studied to reduce the defect generation such as growth on misoriented Si substrates [36], growth on Si-based compliant substrates [37], selective-area growth on patterned substrates [37] and defect gettering via reticulated structures [38], which however received less attention, and thus will not be detailed here.

In all, Si substrate is the most intensely studied alternative substrate for growing CdTe and HgCdTe materials in comparison to other alternative substrates. Significant progress has been made in the epitaxial growth of CdTe and HgCdTe on Si, and the dislocations density (in terms of EPD) in HgCdTe/CdTe/Si are typically $\geqslant$mid-10^6 cm^{-2}. HgCdTe materials grown on Si have been applied for fabricating high-performance SWIR and MWIR detectors. However, there is little information on their applications in LWIR HgCdTe detectors as the EPD level is not low enough to fabricate high-performance LWIR detectors. More detailed discussion on their device application will be introduced in chapter 5. Thus, more effort and novel growth methods are needed to further reduce the dislocation density in CdTe and HgCdTe layers grown and thus improve the material quality and consequently device performance.

3.3 Heteroepitaxial growth of CdTe and HgCdTe on Ge substrates

Ge substrate is a type of mature substrate of higher crystal quality, lower cost, and wide commercially availability compared with traditional CdZnTe substrates. As discussed in chapter 1, Ge substrate presents a smaller lattice and CTE mismatch with HgCdTe in comparison to Si substate, which, in principle, should lead to better material quality. In addition, Ge substrate has the benefit of easy chemical preparation, and the oxide can be desorbed at a lower temperature compared with Si substrate [39]. Therefore, Ge substrate offers a good alternative for growing CdTe and HgCdTe epilayers. So far, MBE growth of HgCdTe/CdTe on Ge substrate has mainly been reported by groups from France and China.

The CEA-Leti group from France has studied the heteroepitaxy of CdTe on several Ge crystalline orientations: misoriented (100), misoriented (111), (331), and (211) [39, 40]. The best results have been obtained with CdTe (211) epitaxial layers on Ge (211) in terms of growth conditions, surface morphology, and crystalline quality. Therefore, this book will mainly focus on the discussion of growing CdTe and HgCdTe layers on Ge (211) substrate. In addition, MBE growth of CdTe and HgCdTe on Ge (211) substrate suffers the challenge of crystallographic polarity, and the crystallographic polarity has to be controlled on the B face (Tellurium) since the MBE growth on the CdTe A face (cadmium) induces many defects [39]. The typical MBE growth process on Ge is as follows [39]:

First, the Ge substrate surface is etched in a chemical solution (H_3PO_4:H_2O_2:H_2O), rinsed in running deionized water and dried with nitrogen gas. Then the Ge wafer is rapidly loaded into the MBE ultra-high vacuum chamber for thermal desorption of native oxide and epilayer growth. The *in-situ* preparation is similar to that of Si alternative substrate. The preparation of Ge surface begins with an outgassing process conducted at 650 °C in the presence of an arsenic flux. This step is essential for cleaning the surface and preparing it for epitaxial growth. Following this, a monolayer of zinc is deposited at a temperature of 280 °C. This zinc layer plays a crucial role in stabilizing and preserving the correct crystalline polarity of the substrate—specifically the B face, which is favorable for the subsequent CdTe growth. Once the surface is properly prepared, an initial buffer layer of CdTe, approximately 500 Å thick, is grown. This initial buffer serves as a foundation for the main CdTe layer, which is typically around 7 μm thick, and is grown at higher temperature (340 °C) to optimize the crystalline quality.

If not performed in a dual chamber MBE growth facility, the CdTe/Ge substrate grown will need to be removed from the chamber growing CdTe, and then chemical prepared and loaded into the HgCdTe growth chamber. For chemical preparation, the substrate surface of CdTe/Ge is etched in a standard bromine and methanol solution used for CdZnTe substrate. Then the CdTe/Ge substrate is reloaded into the MBE chamber for HgCdTe growth. Similar to the growth of HgCdTe on CdZnTe, the oxidized surface is outgassed at 340 °C before growing HgCdTe at about 180 °C. In addition, the growth conditions will need to be well controlled to ensure high-quality HgCdTe growth—especially growth temperature—as the optimum temperature window for growing high-quality HgCdTe is very narrow (only 5 °C). The CEA-Leti group has achieved a temperature stability of ±0.1 °C during growth and a run-to-run accuracy within ±1 °C. It is critical to carefully regulate and monitor both the substrate temperature and the mercury flux, as any deviation can lead to a decline in crystalline quality or an increase in defect density.

With the above growth procedure, high-quality CdTe and HgCdTe layers have been achieved on Ge alternative substrates. A very smooth and mirror-like HgCdTe was obtained on 2-, 3- and 4-inch Ge substrates at CEA-Leti [41, 42]. The EPD observed in HgCdTe grown on Ge substrates, using the Hännert and Schenk etching solution, ranged from 5×10^6 cm^{-2} to 2×10^7 cm^{-2} [41, 42]. These values are similar to those grown on Si, but still significantly higher than those counterpart values on CdZnTe substrates (between 10^4 cm^{-2} and low-10^5 cm^{-2} range). An average XRD FWHM of 80–100 arcsec is typically achieved on CdTe/Ge layers with a thickness of around 7 μm, while values as low as 50 arcsec are sometimes observed on layers that are 10 mm thick [41, 42]. For HgCdTe grown on Ge, XRD FWHM values in the range of 90–130 arcsec are usually achieved [41], indicating high crystalline quality of the epitaxial layers.

So far, most of the study on growing CdTe and HgCdTe on Ge alternative substrates is reported from the group at CEA Leti, France. Most of their studies focus on MWIR HgCdTe materials. HgCdTe IR detectors with different format sizes have been fabricated including 256 × 256 [43], and 1K × 1K [41], which presents excellent device performance. However, as stated above, the dislocation density in HgCdTe grown on Ge ranges from around mid-10^6 cm^{-2} to low-10^7 cm^{-2} which makes it unsuitable for making high-performance LWIR detectors. Although there is a report from Raytheon [44] on growing LWIR HgCdTe on Ge with excellent results (XRD FWHM of 77 arcsec, and EPD of mid-10^6 cm^{-2}), there are no following reports regarding their detector applications. Therefore, more effort is needed to further reduce the dislocation density in CdTe and HgCdTe layers grown and thus improve the material quality and consequently the device performance. In particular, the dislocation reduction techniques discussed in section 3.2 can be applied to improve the material quality and reduce the dislocation density in the CdTe and HgCdTe epilayers grown on Ge, which remain to be studied in the future.

3.4 Heteroepitaxial growth of CdTe and HgCdTe on GaAs substrates

Apart from Si and Ge, GaAs alternative substrate was also intensively studied for growing CdTe and HgCdTe layers. GaAs substrate is also a mature substrate which is of higher crystal quality, lower cost, and wide commercial availability compared with the traditional CdZnTe substrates. As discussed in chapter 1, GaAs substrate also presents a smaller crystal lattice (14.4%) and CTE (14%) mismatch with HgCdTe in comparison to Si substate, which provides a good candidate substrate for growing CdTe and HgCdTe. There are two different growth methods for growing CdTe and HgCdTe on GaAs: (1) MBE growth. For MBE growth, people usually use GaAs substrates with (211) crystal orientation due to the higher Hg sticking coefficient for [211] surface during MBE growth; (2) MOCVD growth. For MOCVD growth, people usually use GaAs substrates with (001) crystal orientation due to the higher Hg sticking coefficient for [001] surface during MOCVD growth. MOCVD growth of CdTe and HgCdTe on GaAs was mainly reported by groups from the United States, United Kingdom, Poland, Japan, and South Korea, while MBE growth of CdTe and HgCdTe on GaAs was mainly reported by groups from Germany, the United States, China, India, Russia, and Turkey. So far, commercial HgCdTe IR detectors grown on GaAs are available from Germany (AIM, MBE growth), the United Kingdom (Selex, MOCVD), Poland (Vigo, MOCVD) as shown on their webpages. As this book mainly focuses on the MBE growth of CdTe and HgCdTe, the following discussion on GaAs alternative substrates will also be mainly on their MBE growths, while some useful knowledge on MOCVD growth will be included in the discussion as well.

As discussed in previous sections, Si and Ge are nonpolar materials, which means their (211) orientation lacks distinct A or B faces. So, their crystal polarity must be defined during the initial stages of buffer layer growth. As a result, arsenic surface termination and ZnTe nucleation layers are often used to define the favorable B face for growing subsequent CdTe and HgCdTe. In contrast, GaAs is a polar crystal, and (211)-oriented GaAs substrates contain both A (Ga-terminated) and B (As-terminated) faces. This allows for the preferred B face polarity required for HgCdTe growth to be obtained simply by selecting the B face of a GaAs wafer, eliminating the need for a specialized nucleation layer [10]. This simplifies the whole MBE growth process. The typical MBE growth process on GaAs is as follows [45]:

Epi-ready GaAs substrates are loaded into the MBE chamber without prior treatment and heated to 580 °C to remove the native oxide layer, under an As overpressure. After the oxide desorption, the substrates are then cooled for a low-temperature (~220 °C) CdTe nucleation step followed by an anneal. The low temperature nucleation is to preserve the (211) growth orientation and potentially to 'bury' residual GaAs surface defects under a thin amorphous-like layer. With RHEED, it can be observed that the crystallinity of this nucleation layer (~250 nm thick) gradually improves during growth and after a high-temperature anneal. This creates a better surface for subsequent CdTe deposition at normal growth temperatures. Then, a thick CdTe buffer layer is grown at a proper temperature (typically between 280 °C and 320 °C) under Te-stabilized growth conditions. High-temperature (~400 °C) *in-situ* annealing (not cyclic annealing) is also undertaken after the growth of a thick CdTe layer by some groups to enhance the crystal quality and annihilate dislocations [46–48] Similar to other alternative substrates, typically a thick CdTe buffer layer is used to reduce the threading dislocations and thus EPD within the materials. Figure 3.1(b) shows the XRD FWHM of the CdTe buffer layers as a function of the CdTe buffer layer thickness when grown on GaAs [11]. It can be observed that a CdTe buffer layer with an appropriate thickness (>10 μm) can effectively block the penetration of dislocations, leading to high crystal quality of the CdTe buffer layer grown on GaAs (XRD FWHM ~60 arcsec). In the meantime, it is also observed that the EPD also decreases with increasing CdTe buffer layer thickness, and the EPD saturates around low-10^6 cm^{-2} to mid-10^6 cm^{-2} when the CdTe buffer layer thickness is over 10 μm. Therefore, a thick CdTe buffer layer (>10 μm) is needed to reduce the dislocation density in the CdTe layer and the subsequent HgCdTe epilayers [11]. Apart from using a thick CdTe layer, the growth conditions/parameters can also be studied and optimized to improve the material quality and reduce the dislocation density within. These growth conditions/parameters include: VI/II flux ratio [49], detailed growth temperature profile [50], annealing temperature [47, 48], and others.

In comparison to other alternative substrates, GaAs was most intensively studied in 1990s, and various dislocation reduction approaches were investigated to improve the material crystal quality and reduce the dislocation density in the CdTe and HgCdTe layers. The following are some dislocation reduction techniques reported for GaAs alternative substrates:

In-situ cyclic annealing: As reported by Jacobs *et al* [45], *in-situ* cyclic annealing was introduced during the MBE growth of a thick CdTe layer (after the low-temperature CdTe nucleation and anneal) to promote the interaction and elimination of threading dislocations. In their work [45], each typical cycle involves depositing 1 μm of CdTe, followed by a 5 min anneal under Te-rich conditions. Temperature ramping rates are generally around 0.5 °C min^{-1}. The CdTe layer thickness usually falls within the range of 9–13 μm. Following the growth of the CdTe buffer layer, the CdTe/GaAs composite substrate was transferred under vacuum to a second MBE chamber for the deposition of HgCdTe which was grown via a routine growth process using Hg, Cd, or CdTe, and Te fluxes and at proper growth temperature.

The above procedure for CdTe buffer layer growth was very successful, and 3-inch CdTe/GaAs wafers were obtained with uniform and smooth surfaces (RMS (root mean square) surface roughness of 1.4 nm) [45]. Figure 3.9 shows the XRD FWHM map of an optimized 3-inch CdTe/GaAs wafer [45]. It is observed that a uniform XRD FWHM of 68 arcsec is measured throughout most of the CdTe buffer layer. Deviations from this average value are typically observed toward the edge of the wafer, which can be partially due to the effect of wafer bowing. For typical 10-μm-thick layers, relatively uniform 60 arcsec is observed across the full 3-inch wafer with the lowest value of 52 arcsec as reported by Jacobs *et al* [45], which indicates excellent crystalline quality for the CdTe/GaAs wafers. The cathodoluminescence (CL) imaging on the CdTe/GaAs wafers shows that the threading dislocation densities are in the range of high-10^6 cm^2 to low-10^7 cm^2 for the samples examined. Note that CL imaging measurements are more likely to overcount (rather than undercount) the number of threading dislocations due to the effective probe depth of the technique and existence of subsurface defects [45].

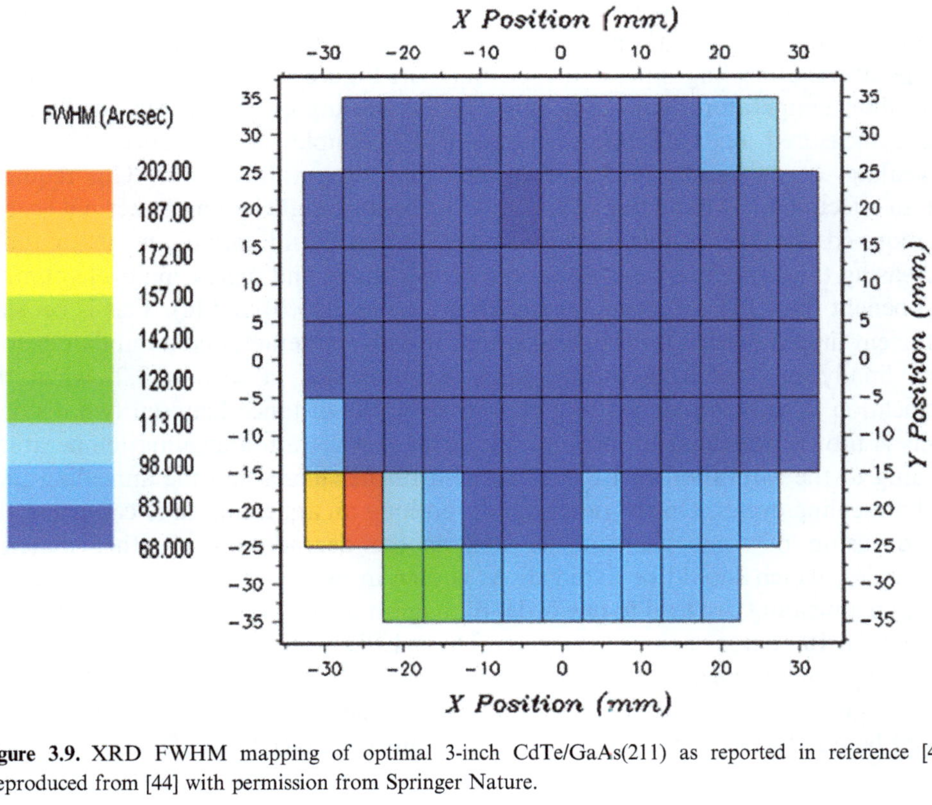

Figure 3.9. XRD FWHM mapping of optimal 3-inch CdTe/GaAs(211) as reported in reference [44]. Reproduced from [44] with permission from Springer Nature.

Apart from high-quality CdTe/GaAs layers, the HgCdTe epilayers grown subsequently also show high material quality with relatively smooth surfaces. For LWIR $Hg_{1-x}Cd_xTe$ ($x = 0.23$) layers grown, the as-grown surface defect densities are typically 1000 cm^2 and primarily consisted of discrete voids and tellurium precipitates. Optimized HgCdTe layers have an EPD as low as 2×10^6 cm^{-2}. At 80 K, the n-type LWIR HgCdTe layers grown show a typical n-type carrier concentration of 2×10^{15} cm^3 and an electron mobility 1×10^5 cm^2 V^{-1} s^{-1}, respectively. The best values achieved for 80 K carrier concentration and electron mobility are 1.4×10^{15} cm^3 and 126 000 cm^2 V^{-1} s^{-1} [45]. Such a high electron mobility is comparable to that of LWIR HgCdTe grown on lattice-matched CdZnTe substrates [3].

Postgrowth cyclic annealing: Apart from *in-situ* cyclic thermal annealing, post-growth cyclic thermal annealing was more intensively studied to improve the material quality and reduce the dislocation density in CdTe and HgCdTe layers grown on GaAs substrates [45, 51–53].

Bakali *et al* [53] reported a study on post-growth cyclic thermal annealing of CdTe layers grown on GaAs. In their work, the heater temperature was kept constant for a certain annealing duration. Cyclic annealing was carried out by heating a sample rapidly up to a set annealing temperature at which the sample was annealed for a fixed duration, then the heater current was turned off and the sample was naturally cooled for 15 min in each cycle. This process was repeated for a number of cycles. The total annealing time was calculated as the number of cycles multiplied by the annealing duration of each cycle. And, the influence of annealing conditions on the ultimate annealing results were investigated such as annealing temperature, annealing time, and cycle number. Figure 3.10 shows the influence of annealing temperature, annealing time, and annealing cycle number on the EPD value measured on CdTe/GaAs layers (CT9 samples in reference [53]) after annealing. It can be observed that an annealing temperature $\geq$400 °C is required for an effective EPD reduction. This might be because sufficient high thermal energy is required for the dislocations to move, interact and annihilate. Also, both increasing the annealing time within the proper range and increasing cycle number can benefit the EPD reduction, as shown in figures 3.10(b) and (c). This is because sufficient time is needed for the dislocations to move, interact, and annihilate before they finally saturate at a certain level. As discussed in section 3.2, when the dislocation density is reduced to a certain level, the distance between two dislocations is too far for them to move and interact at a certain annealing temperature, leading to the saturation of EPD value with further increasing the annealing time and annealing cycles. On the other hand, too long an annealing time can cause the out-diffusion of some elements as well as the degradation of other material properties, which should be avoided. As shown in figure 3.10, generally the cyclic thermal annealing can lead to one order of magnitude reduction in the EPD, namely from high-10^7 cm^{-2} to low-10^7 cm^{-2}. These EPD results both before and after annealing are not ideal in terms of material quality, but they do show the feasibility of post-growth cyclic annealing for EDP reduction. The high initial EPD values might be due to the non-ideal growth conditions for CdTe/GaAs layers.

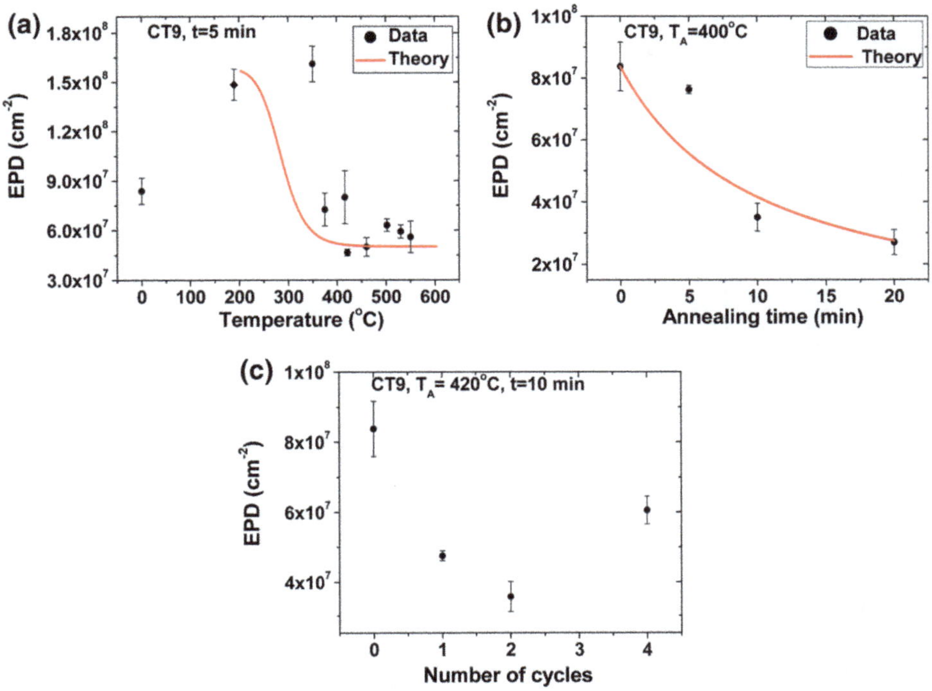

Figure 3.10. Impact of post-growth cyclic annealing conditions on the EPD of CdTe layers grown on GaAs as reported in reference [53]. Reproduced from [53] with permission from Springer Nature.

Apart from CdTe/GaAs layers, post-growth cyclic annealing was also studied for reducing the dislocation density (in terms of EPD) in HgCdTe layers grown on CdTe/GaAs wafers. Shin *et al* [52], from Rockwell reported an interesting study on applying both normal and cyclic post-growth thermal annealing to reduce the dislocation density in LWIR $Hg_{1-x}Cd_xTe$ ($x = 0.22$) grown on CdTe/GaAs. Initially, they just undertook normal thermal annealing to study its impact on dislocation reduction. In this case, the HgCdTe/CdTe/GaAs epilayers were annealed at high temperatures (410 °C–490 °C) in a high-pressure furnace with Hg overpressure to reduce the dislocation density within. After the high-temperature annealing, the epilayers were n-type annealed at 250 °C under an Hg environment to remove nonstoichiometric Hg vacancies induced by the high-temperature annealing process. The threading dislocation densities were characterized by measuring EPD revealed by chemical etching using the $HNO_3/K_2Cr_2O_7/HCl/H_2O$ chemical solution [52]. Typically, one order of magnitude reduction in dislocation density was observed in terms of EPD values. Figure 3.11 presents the result of EPD reduction in HgCdTe layers after thermal annealing. It can be observed that the EPD is 7×10^6 cm^{-2} before thermal annealing, but 8×10^5 cm^{-2} after thermal annealing.

Based on the dislocation density reduction results of normal thermal annealing, further reduction of the dislocation density in MBE-grown HgCdTe layers on GaAs substrates can be achieved by applying cyclic thermal annealing. The following shows an example cyclic thermal annealing process reported by Rockwell group [52]:

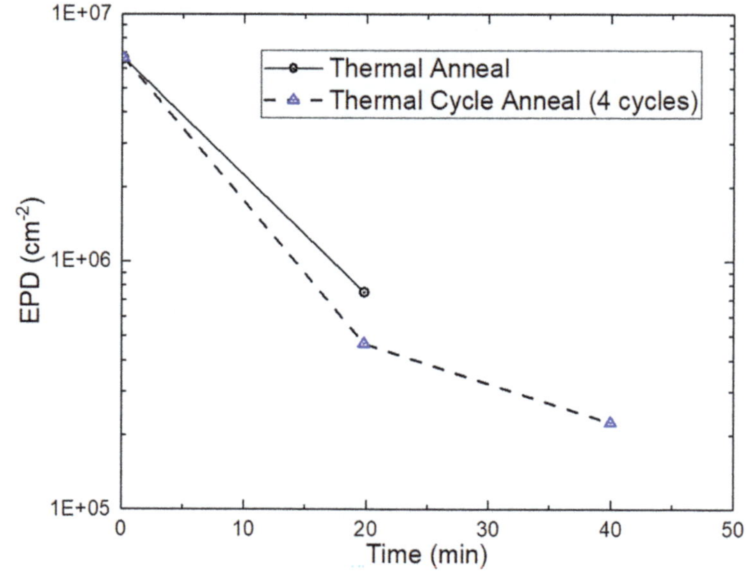

Figure 3.11. EPD dependence on the number of repetitions at 490 °C–300 °C cyclic thermal annealing and single thermal cyclic annealing for the sample periods of time. Reproduced from [52] with the permission of AIP Publishing.

(1) first the as-grown MBE HgCdTe film with an EPD of 7.7×10^6 cm^{-2} was loaded in a high-pressure furnace with Hg, and (2) the sample was annealed for 3–5 min at 490 °C under H_2 overpressure, followed by lowering the furnace temperature to 300 °C. This cyclic annealing between 490 °C and 300 °C was repeated four times in each annealing run. In the end, the Rockwell group demonstrated high-quality HgCdTe films with an EPD of 2.3×10^5 cm^{-2}, which is a factor of two to three lower than a single high-temperature annealing result. As stated before, an EPD value of 6.9×10^6 cm^{-2} was measured on the as-grown HgCdTe/CdTe/GaAs material, but after the single thermal annealing was carried out, the surface EPD value dropped by an order of magnitude to 7.7×10^5 cm^{-2}. However, after two runs of the cyclic thermal annealing process, the surface EPD value dropped further to 2.3×10^5 cm^{-2}. Figure 3.11 summarizes the results of normal thermal annealing and cyclic thermal annealing of MBE-grown HgCdTe/CdTe/GaAs materials. These results suggest that thermal annealing (especially cyclic thermal annealing) provides an effective way to reduce the dislocation density in HgCdTe layers grown on GaAs in proper annealing conditions.

Apart from dislocation density, the Rockwell group also studied the relationship between minority carrier lifetime and EPD in n-type LWIR HgCdTe layers grown on CdZnTe and CdTe/GaAs substrates [52]. It was found that with the same electron concentration the minority carrier lifetime decreased with increasing EPD value once the EPD was above the level of 5×10^5 cm^{-2}. However, the minority carrier lifetime maintained an almost constant value if the EPD was below the level of 5×10^5 cm^{-2}. Based on this, Rockwell proposed the criteria EPD level of 5×10^5 cm^{-2} for making high-performance LWIR detectors.

As discussed above, thermal annealing is one of the most effective processes for annihilating threading dislocations because it increases dislocation motion with subsequent interaction, annihilation, and formation of looping networks which effectively pin threading dislocations. The reduction of dislocations in MBE-grown HgCdTe/GaAs could probably occur by pinning dislocations at Hg-vacancies point defects formed during growth and high temperature annealing [46]. As reported by the Rockwell group [52], the dislocation density in HgCdTe layers grown on GaAs could be reduced from 7×10^6 cm^{-2} to as low as 2.3×10^5 cm^{-2} by cyclic thermal annealing between 300 °C and 490 °C, which is comparable to the values obtained for growth on CdZnTe. Although this post-growth annealing is very effective in reducing the dislocation density and improving the material quality, it suffers a major limitation due to interdiffusion of elements, which degrades the sharpness of hetero- and homojunctions required for advanced detector designs.

Strained superlattice dislocation filter: Apart from *in-situ* thermal annealing and post-growth thermal annealing, strained superlattice layers provide another effective approach to reduce the dislocation density in CdTe and HgCdTe layers grown on GaAs alternative substrates. Sugiyama *et al* presented a study on using CdTe/$Cd_{0.93}Zn_{0.07}Te$ strained superlattice layers to block the threading dislocations extending from the buffer/substrate interface during the hot-wall epitaxy of CdTe on GaAs (001) substrates [54]. The CdTe/GaAs wafers grown with superlattice layers were used for the subsequent MOCVD growth of HgCdTe on its top. Sugiyama *et al* devoted significant effort to designing the thickness of CdTe and CdZnTe layers constituting the superlattices to ensure the superlattices having a layer force greater than the residual compressive force, but less than the generation range of threading dislocations. By introducing two sets of strained superlattice layers with properly designed layer thickness for CdTe and CdZnTe layers, the EPD in CdTe layers was reduced from 2.8×10^8 cm^{-2} to 8.2×10^7 cm^{-2}, and the subsequent EPD in HgCdTe layer was reduced from 2.3×10^6 cm^{-2} to 1.1×10^6 cm^{-2}. In the meantime, the XRD FWHM value was also reduced from 185 to 124 arcsec. Although the CdTe/$Cd_{0.93}Zn_{0.07}Te$ strained superlattice layers were grown with hot wall epitaxy, the results can still indicate the feasibility of using strained superlattice layers for blocking and filtering threading dislocations within the CdTe and HgCdTe layers.

Most recently, the research group at UWA (the author's group of this book) undertook a detailed study on using CdTe/CdZnTe strained superlattice layers as dislocation filtering layers to block and reduce the threading dislocations in the CdTe layers grown on GaAs (211)B substrates with MBE growth [55, 56]. For the convenience of later discussion, dislocation filtering layers are abbreviated as 'DFLs.' Note that in Sugiyama's study on hot wall epitaxy of CdTe/CdZnTe superlattices [54], the chemical composition of CdZnTe could not be changed, and thus Sugiyama *et al* had to adjust the layer thickness to design and control the strain between layers. Instead, with the MBE growth technique, CdZnTe layers with various chemical compositions could be grown which provides another parameter to design and tune the strain between layers. In addition, *in-situ* thermal annealing under a Te environment is also available with MBE growth which can be applied to further facilitate the movement and interaction of dislocations and thus the annihilation of threading dislocations. The reported MBE growth process of CdTe buffer layer on GaAs with the incorporation of CdTe/CdZnTe strained superlattice layers at UWA is as follows [55]:

CdTe buffer layers were grown on 2-inch GaAs (211)B substrates. Elementary Zn, Te, and compound CdTe were used as MBE source materials for growing CdTe layers. After oxide desorption at ~580 °C for 3 min, the substrates were cooled down to 220 °C. Then a thin ZnTe nucleation layer (<30 nm) was grown at 220 °C, and then *in-situ* thermal annealed with a background Te beam equivalent pressure (BEP) of 3×10^{-6} Torr at 380 °C for 15 min. The purpose of this thin ZnTe nucleation layer is to suppress three-dimensional growth and preserve a good (211)B interface for the subsequent CdTe growth [57]. After growth of a 2–5 μm thick CdTe bottom layer, several sets of CdZnTe/CdTe strained superlattice layers were grown, separated by 500-nm-thick CdTe spacer layers. Each set of CdZnTe/CdTe strained superlattice layers consisted of five periods of ~13 nm CdZnTe/~11 nm CdTe. *In-situ* thermal annealing for 15 min was undertaken after the growth of each set of superlattice layers. The samples were completed by the growth of 4–20 μm thick CdTe top layer. Both the CdZnTe and CdTe layers were grown at about 280 °C at a growth rate of ~2 $\mu m\ h^{-1}$ under Te-rich conditions with the Te/CdTe BEP ratio of approximately 1.5.

To better understand the effect of strained superlattice layers on dislocation density reduction, the UWA group studied different superlattice structural designs and annealing conditions. Figure 3.12(a) shows a typical schematic sample structure for a CdTe buffer layer on a 2-inch GaAs (211)B substrate with five sets of $Cd_{1-x}Zn_xTe$/ CdTe superlattice layers. The purpose of this superlattice layer structure is to increase the likelihood of dislocation annihilation and inhibit dislocation multiplication and/or propagation. As discussed by Ward *et al* [58] the key to significant reduction in threading dislocations is the dislcoation movement using misfit stress, i.e., in layers with a finite strain-thickness product (εh). Moreover, a good rule of thumb for DFL design in heteroepitaxial growth is given by $\varepsilon_c h_c < \varepsilon h < 4\varepsilon_c h_c$, where $\varepsilon_c h_c$ is a constant (experimentally, $\varepsilon_c h_c$ is ~0.24 nm in the GaInAs system) [59]. The constraint condition determines a limited Zn content range of $0.08 < x < 0.30$ for the superlattice layer design, assuming $\varepsilon_c h_c = 0.24$ nm for the $Cd_{1-x}Zn_xTe$ system. Note that a high-quality CdZnTe buffer is challenging to grow, and layers that are too thick and/or have more Zn content may increase the risk of Zn segregation. However, since the CdZnTe superlattice layers here are only of the order of 10 nm thick, subsequent CdTe layers should smooth out any irregularities in the CdZnTe surfaces before other crystal phases nucleate. Figure 3.12(b) presents bright-field (BF) cross-sectional transmission electron microscopy (TEM) images of CdTe buffer layers grown on GaAs (211)B with CdZnTe/CdTe strained superlattice layers, showing clear and sharp interfaces between the superlattice layers and the surrounding CdTe layers. In addition to sharp superlattice interfaces, apparent redirection and termination of threading dislocations are observed at the CdZnTe/ CdTe hetero-interfaces, as indicated by the yellow arrows.

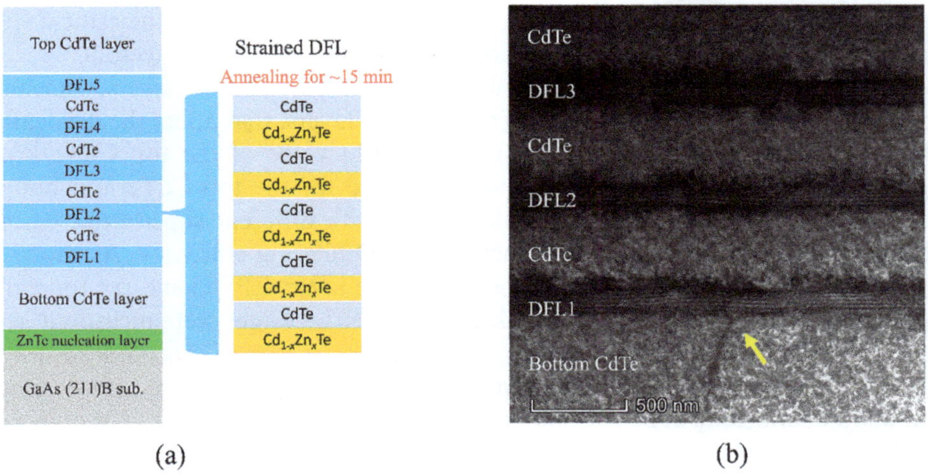

Figure 3.12. (a) Schematic sample structure for CdTe buffer layer on GaAs (211)B with several sets of strained CdZnTe/CdTe superlattice dislocation filtering layers. Each set of superlattice layers consists of 5 periods of $Cd_{1-x}Zn_xTe$/CdTe with thicknesses of ~13 nm/~11 nm, respectively; (b) TEM image of CdTe buffer layer grown on GaAs (211)B substrate with CdZnTe/CdTe superlattice layers, showing evidence for dislocation filtering (indicated by yellow arrows). Reproduced from [55] with permission from Springer Nature.

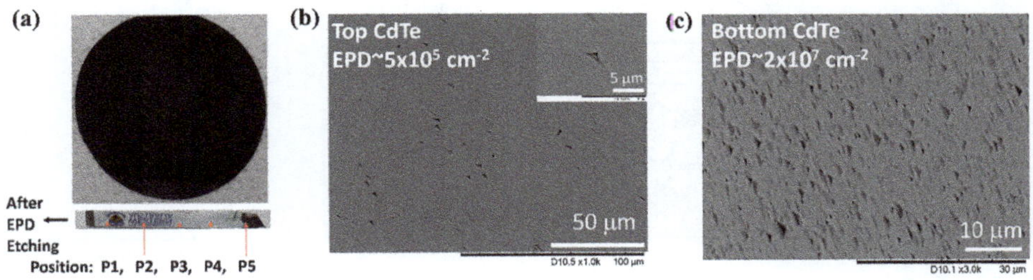

Figure 3.13. (a) Photograph of 2-inch MBE-grown CdTe buffer/GaAs (211)B with DFL annealed at 320 °C; (b) SEM image for EPD measurement of top CdTe layer; (c) SEM image for EPD measurement for the bottom CdTe surface after EPD etching. Reproduced from [55] with permission from Springer Nature.

At UWA, most CdTe buffer layers were grown with a sandwich structure of ~5 μm CdTe bottom layer/superlattice layer region (~2.8 μm thick for five sets of superlattice layers)/~5 μm CdTe top layer. Figure 3.13(a) shows an MBE-grown 2-inch CdTe buffer/GaAs (211)B sample, having a uniform and mirror-like surface to the naked eye. In order to examine the material quality of the top CdTe layer across the whole wafer, EPD measurements were performed for a bar cut along the diameter, which still shows a shiny surface after EPD etching, as visible in figure 3.13 (a). Figure 3.13(b) shows a representative SEM image for the etched surface, and the EPD is determined to be approximately 5×10^5 cm^{-2}, agreeing with values determined from optical imaging. It should be mentioned that the EPD value is highly uniform along the wafer diameter. In contrast, as evident from figure 3.13(c), the EPD of the bottom CdTe layer is around 2×10^7 cm^{-2}, suggesting that the superlattice dislocation filtering layers reduces the dislocation density by a factor of

40. The EPD level achieved for the top CdTe layer is approaching the critical EPD level required for fabricating high-performance LWIR HgCdTe detectors.

There are several parameters such as x value in $Cd_{1-x}Zn_xTe$ layers, annealing temperature, and the number of sets of superlattice layers, that affect the dislocation filtering efficiency (η) and therefore the final EPD value of the top CdTe layer. Figure 3.14 summarizes the EPD values measured for the top CdTe and bottom CdTe layers of the CdTe/GaAs samples grown with different superlattice designs (different Zn content (x)), and corresponding EPD filtering efficiency. As shown in figure 3.14(a), the EPD values of the bottom CdTe layer (EPD_b) and the top CdTe layer (EPD_t) in CdTe buffer layers with different DFL designs were measured, with the corresponding η as calculated by $\eta = (EPD_b - EPD_t)/EPD_b$ and shown in figure 3.14(b). In the case of buffer layers with five sets of superlattice layers grown with thermal annealng at 320 °C, an optimal x range of $0.15 < x < 0.17$ is observed, which leads to a low EPD level ranging from mid-10^5 to low-10^6 cm^{-2}, and the highest EPD filtering efficiency of ~98% within the Zn composition range of $0.08 < x < 0.27$ studied. When $x = 0.17$, samples with no thermal annealing as well as those that underwent annealing at the higher temperature of 360 °C result in a relative

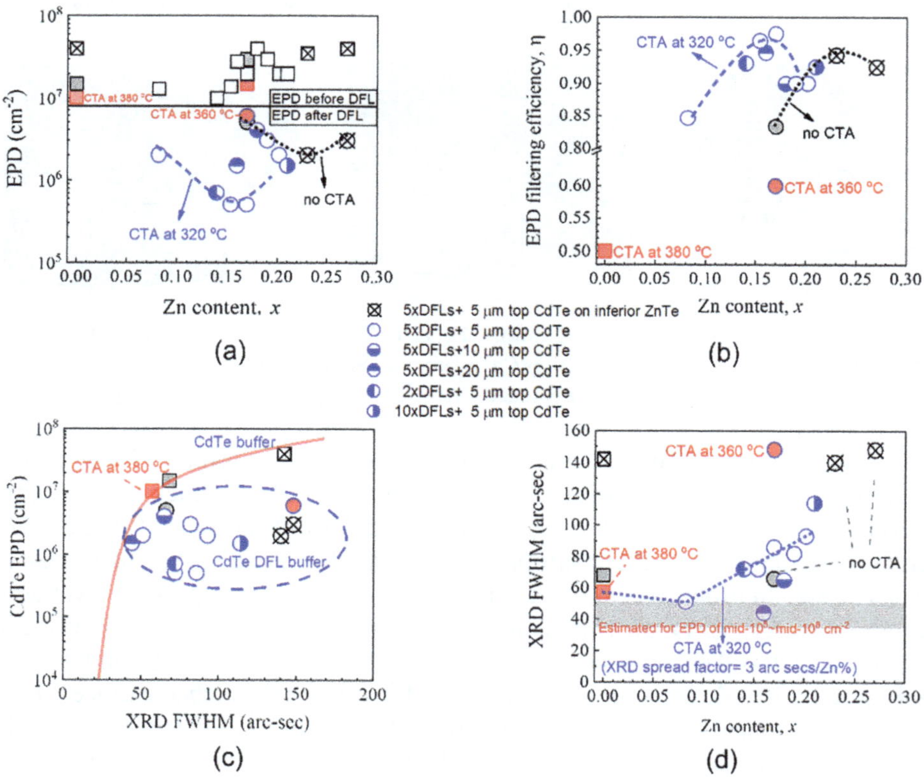

Figure 3.14. (a) EPD values measured for the top CdTe and bottom CdTe layers for DFL buffer layers grown with different DFL designs and plotted as a function of Zn content (x), and (b) corresponding EPD filtering efficiency. Note "CTA" refers to the in-situ thermal annealing undertaken. Reproduced from [55] with permission from Springer Nature.

higher EPD level of mid-10^6 cm^{-2} and corresponding η of 85% and 60%, respectively. It is generally assumed that moderate thermal annealing for superlattice dislocation filtering structures can further improve their efficiency of filtering defects due to the combined effect of increasing the mobility of defects and introducing thermal strain (ε_t) for more efficient defect annihilation [60]. Between room temperature and the highest thermal annealing temperature of 360 °C used, this difference in thermal expansion amounts to $\varepsilon_t = 2.4 \times 10^{-4}$, which is nearly two orders of magnitude lower than the lattice-mismatch-induced strain of $\varepsilon = 1 \times 10^{-2}$ in a $Cd_{1-x}Zn_xTe/CdTe$ superlattice layer with $x = 0.17$. Therefore, thermal mismatch appears to play a minor role in the dislocation reduction process for the CdTe buffer layers on GaAs with CdZnTe/CdTe superlattice layers and thermal annealing at the temperatures evaluated (<360 °C). It was reported that for many lattice-mismatched systems such as strained InGaAs/GaAs and strained SiGe/Ge, annealing of the strained layers helps to generate misfit dislocations that relax the strain and reduce critical thickness [61–63]. Therefore, the decrease of dislocation filtering efficiency observed in the CdTe buffer with superlattice layers annealed at 360 °C could be attributed to thermal-annealing-assisted misfit dislocation formation, and strain relaxation/leakage in the CdZnTe layers. Further fine control of the annealing temperature and/or annealing duration, as well as Te flux, could be effective in achieving CdTe buffers with EPD_t < mid-10^5 cm^{-2}.

As evident from figure 3.14(b), other parameters such as thicknesses of the top and bottom CdTe layers and the number of sets of strained superlattice layers, are also expected to affect the EPD_t values. Therefore, further reduction of EPD in the CdTe/GaAs wafer to below the critical level of 5×10^5 cm^{-2} is possible with further optimizing of the relevant parameters of strained superlattice layers. However, it should be noted that it still presents a challenge to grow these CdZnTe/CdTe strained superlattice layers with the correct parameters (compostion, thickness, etc).

In all, there were rather comprehensive studies of both MBE and MOCVD growth of CdTe and HgCdTe layers on GaAs alternative substrates, especially in the 1990s. Various experimental techniques were proposed to reduce the threading dislocation density within, and excellent results such as an EDP of 2.3×10^5 cm^{-2} were already reported for LWIR HgCdTe on CdTe/GaAs with cyclic thermal annealing, which is below the critical EDP level of 5×10^5 cm^{-2} for making high-performance LWIR HgCdTe detectors. However, those dislocation reduction methods are still complex for industry manufacturing, and most of the EPD results after applying dislocation reduction techniques are still higher than the critical EDP level of 5×10^5 cm^{-2} and are unsuitable for making high-performance LWIR detectors. As a result, although there were many research groups/labs working on the growth of CdTe and HgCdTe on GaAs in 1990s, there are only few groups/labs manufacturing HgCdTe IR detectors on GaAs substrates, for example, AIM (Germany), and Shanghai Institute of Technical Physics (SITP, China) with the MBE growth technique, and Selex (United Kingdom), and Vigo (Poland) with the MOCVD growth technique. Also, no manufacturer offers products of LWIR HgCdTe detectors grown on GaAs substrates.

3.5 Heteroepitaxial growth of CdTe and HgCdTe on GaSb substrates

3.5.1 Theoretical perspectives

As discussed in previous sections, alternative substrates (such as Si, GaAs, and Ge) have been intensively studied over the past several decades. However, HgCdTe materials grown on these alternative substrates still suffer a high threading dislocation density (typically >mid-10^6 cm^{-2}) due to the large lattice and CTE mismatch between HgCdTe and the substrates, which makes them unsuitable for making high-performance LWIR detectors. To further enhance the device performance of HgCdTe IR detectors, especially LWIR ones, it is essential to study new alternative substrates better matching HgCdTe materials in terms of lattice constant and CTE. At UWA, we proposed a new GaSb alternative substrate for the MBE growth of high-quality HgCdTe materials for making IR detectors, especially LWIR ones [7, 13, 64, 65]. As shown in figure 1.7, GaSb presents a much smaller lattice constant mismatch with HgCdTe in comparison to Si, Ge, and GaAs. In addition, the CTE mismatch between GaSb and HgCdTe (23%) is also much smaller than that between Si and HgCdTe, and is comparable to that between HgCdTe and both Ge and GaAs. So, theoretically, GaSb provides a better choice as an alternative substrate for the epitaxial growth of HgCdTe/CdTe epitaxial layers for IR detector applications.

Apart from being better matched to HgCdTe, GaSb substrate is also suitable for growing IR materials for industry production. As GaAs alternative substrates are also III–V substrates, and have already been utilized for manufacturing MWIR HgCdTe detectors, here we compare the main wafer characteristics between GaAs and GaSb which will provide a direct indication as to whether or not GaSb alternative substrates are suitable for industry applications. Table 3.1 lists the main characteristics of commercially available GaAs and GaSb substrates [7]. As listed in table 3.1, GaSb substrates have similar IR transmission and commercial availability to GaAs substrates, although GaSb substrates have better material quality in terms of narrower XRD FWHM, as well as lower EPD values. Although GaSb substrate presents a slightly higher cost than GaAs, the smaller lattice mismatch will enable the growth of HgCdTe materials with higher crystal quality

Table 3.1. Main characteristics of GaAs and GaSb as alternative substrates for growing HgCdTe. Reproduced from [7] with permission from Springer Nature.

(211)B wafer	XRD FWHM (arcsec)[1]	EPD (cm^{-2})[2]	IR transmission (%)[3]	Max. diam. (inch)[4]	Unit price (US$ for 2″ wafer)
GaAs	24–30	<5000	48–55	6	150
GaSb	20–25	<3000	47–52	4	550

Notes[1] For omega scan (according to Wafer technology, UK).
[2] From Wafer technology, UK.
[3] For 2∼15 μm wavelength range (according to Wafer technology, UK).
[4] Some manufacturers can provide even large size wafers.

than GaAs can offer. This will not only make low cost and large format LWIR FPAs possible, but also further drive down the cost of MWIR FPAs due to higher FPA yield. Therefore, in practice GaSb substrates have the potential to become a new alternative substrate technology for the growth of high-quality HgCdTe epitaxial layers.

3.5.2 MBE growth of CdTe and HgCdTe on GaSb substrates

As discussed in section 3.5.1, theoretically GaSb presents a better alternative substrate for the MBE growth of HgCdTe due to the smaller lattice mismatch with HgCdTe. In order to study this, MBE growth of CdTe and HgCdTe epilayers were first proposed and undertaken on GaSb (211) B substrates at UWA. In order to relax the lattice mismatched strain and annihilate any misfit and threading dislocations, a thick CdTe buffer layer ($\geq$5 μm) was grown before growing the HgCdTe layer, which is a common procedure used for the MBE growth of HgCdTe on lattice-mismatched substrates as discussed in previous sections. As reported before [10], the dislocation density nucleated in the CdTe buffer layer represents the lower limit on the ultimate dislocation density achievable in the subsequently grown HgCdTe layer. Therefore, significant effort is devoted to growing CdTe buffer layers with the highest material quality. At UWA, we proposed three buffer layer technologies to reduce the dislocation density in the CdTe buffer layers and thus that in the HgCdTe epilayers grown subsequently. These three buffer layer technologies are:

3.5.2.1 Direct CdTe buffer layer technology

For direct CdTe buffer layer technology, a thick CdTe buffer layer ($\geq$5 μm) is directly grown on the lattice mismatched GaSb (211)B substrate before growing the subsequent HgCdTe layer [7]. This thick CdTe buffer layer will benefit the relaxation of the mismatch strain and the annihilation of the misfit and threading dislocations caused by the lattice mismatch. For comparison study, CdTe buffer layers were also grown on GaAs substrates at UWA. As discussed in section 3.5.1, as GaAs substrates have already been used for manufacturing MWIR HgCdTe detectors, a comparison between GaAs and GaSb will provide a direct indication as to whether or not GaSb alternative substrates are feasible for the growth of high-quality HgCdTe materials and the ultimate industry applications.

The reported UWA MBE process for growing CdTe layers directly on GaSb is as follows [7]: epi-ready GaSb (211) B substrates were used as the starting substrates, and high-purity CdTe (6N), Te (6N), and Sb (6N) served as the source materials. After loading the GaSb substrates into the MBE system, the CdTe buffer layers were grown using the following method: initially, the substrate temperature was ramped up to 580 °C to remove the native oxide from the GaSb surface. Subsequently, the temperature was lowered to the desired growth temperature (T_g), and the CdTe buffer layer was deposited at a growth rate of 1 μm h^{-1}. The growth of the CdTe buffer layers began with the simultaneous introduction of CdTe and Te fluxes into the growth chamber. Upon completion of the CdTe layer, the CdTe flux was halted by closing the shutter to the CdTe cell, and the samples were cooled to 260 °C while being maintained under a background Te flux. When the substrate temperature dropped below 260 °C, the Te flux was also stopped by closing the shutter to the Te cell. The samples were then cooled to room temperature without any additional Te flux.

As background Sb flux protection is important for the thermal desorption of native oxide from GaSb substrates, experiments were undertaken using both zero Sb flux and a low Sb flux (BEP of 5×10^{-7} Torr) during the oxide desorption process. To examine how growth temperature and the VI/II flux ratio influence the material quality of CdTe buffer layers, the growth temperature (T_g) was adjusted between 265 °C and 285 °C, while the VI/II flux ratio was varied from 1.5 to 3. For comparison purposes, CdTe buffer layers were also deposited on epi-ready GaAs (211) B substrates, incorporating a thin ZnTe nucleation layer to minimize twin defect formation [66]. During the oxide thermal desorption and buffer layer growth, *in-situ* RHEED was used to monitor the change of surface crystal quality and surface roughness.

Figure 3.15 shows the AFM (atomic force microscopy) image of the GaSb substrate in its as-received state, along with AFM and RHEED images taken after oxide thermal desorption, both with and without a background Sb flux. When no Sb flux is applied, the substrate surface becomes quite rough following oxide desorption, exhibiting an RMS surface roughness of approximately 6.8 nm. This is further confirmed by the elongated dot pattern observed in the RHEED image shown in the inset of figure 3.15(b). However, when a small background flux of Sb (5×10^{-7} Torr BEP) is used during the desorption process, the surface smoothens considerably, with an RMS roughness of around 4.1 nm, as evidenced by the short streak pattern in the RHEED image inset of figure 3.15(c). The significant surface roughness observed after oxide thermal desorption may be attributed to two main factors: (1) insufficient background Sb flux during the desorption process, or (2) poor initial surface quality of the wafers provided by the supplier. According to the wafer supplier, these (211)B GaSb wafers were polished using a method typically applied to (001) oriented GaSb wafers, which may have resulted in a compromised surface quality. This is supported by the AFM image in figure 3.15(a), showing an RMS surface roughness of approximately 2.1 nm for the as-received substrates. Because the UWA MBE chamber is dedicated for II–VI material growth, it is not appropriate to introduce a large Sb flux into the chamber, which limits the protection effect of Sb flux. In the future, atomic or molecular hydrogen cleaning offers an alternative method for removing the native oxide layer from the GaSb substrate surface, which represents a potential area for future investigation [67].

Using the results from oxide thermal desorption with background Sb flux protection, CdTe buffer layers can be grown directly on GaSb substrates without using a ZnTe nucleation layer. To better understand the MBE growth behavior of CdTe on GaSb, two key parameters—growth temperature and VI/II flux ratio—were systematically varied and studied at UWA. Figure 3.16 displays the RHEED patterns and XRD rocking curves for CdTe buffer layers grown at different temperatures. Under a VI/II flux ratio of 2.5, a growth temperature of 275 °C leads to the highest crystal quality for CdTe buffer layers, as indicated by the narrowest

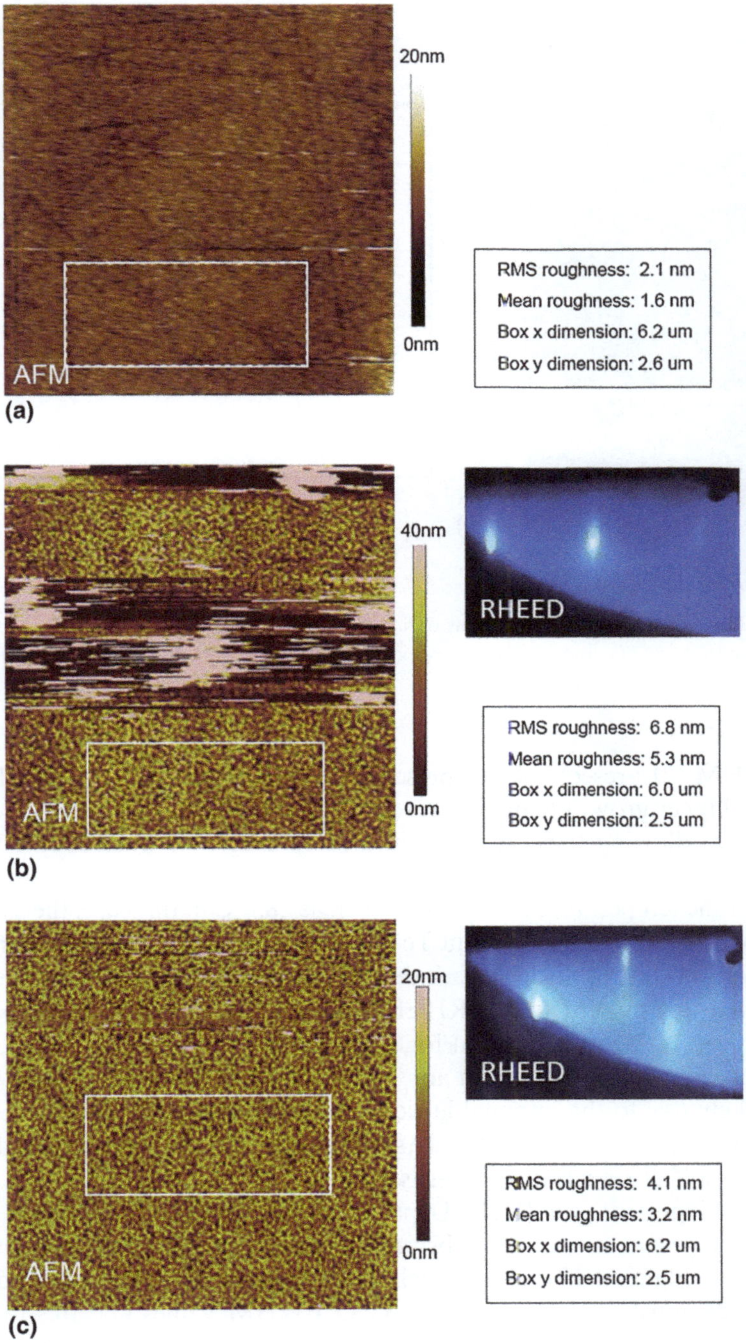

Figure 3.15. Representative AFM image of GaSb substrate surface as-received (a), and AFM images and RHEED patterns of GaSb substrate surface after oxide thermal desorption without (b), and with (c) a background Sb flux protection. The AFM scan size is 10 μm × 10 μm. The white rectangle boxes in the AFM images indicate the areas used for surface roughness measurements. Reproduced from [7] with permission from Springer Nature.

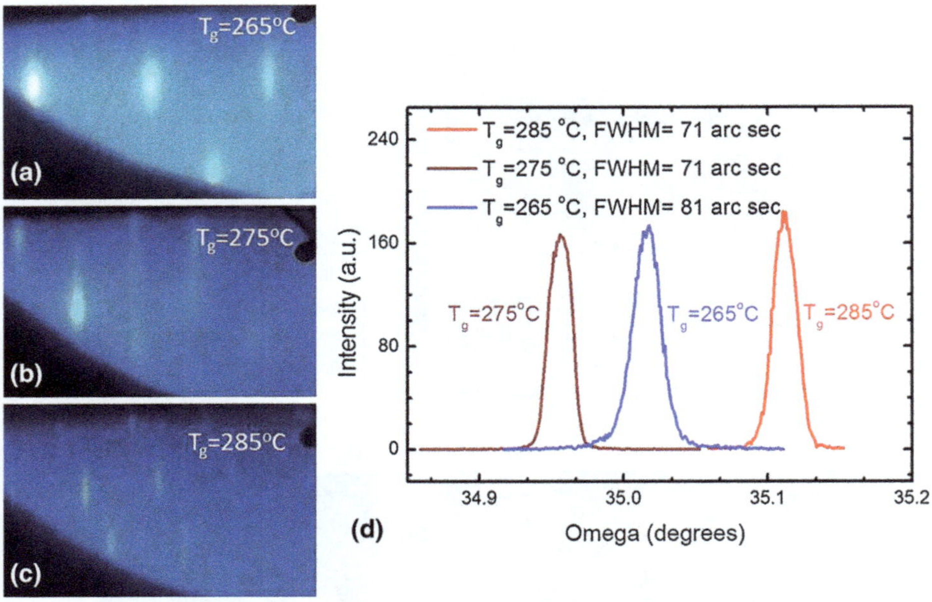

Figure 3.16. Representative RHEED patterns during the growth of CdTe buffer layers at T_g of (a) 265 °C, (b) 275 °C, and (c) 285 °C, and (d) their corresponding XRD rocking curves. Reproduced from [7] with permission from Springer Nature.

XRD FWHM (71 arcsec) and the presence of long streaks in the RHEED pattern. In general, higher growth temperatures (T_g) provide atoms at the growth front with more energy, allowing them to settle into more favorable nucleation sites, which promotes the formation of single-crystal material with improved crystalline quality. However, if the temperature is too high—such as 285 °C in this case, the available Te atoms may be insufficient to prevent Te desorption from the surface under the given VI/II flux ratio. This results in surface degradation, as reflected by the appearance of an elongated dot pattern in the RHEED image. Thus, a growth temperature of 275 °C appears to be near optimal for a VI/II flux ratio of 2.5. Notably, RHEED analysis reveals no twin defects in any of the CdTe buffer layers grown on GaSb, which is likely due to the minimal lattice mismatch between CdTe and GaSb. This contrasts with CdTe growth on GaAs substrates, where twin defects frequently occur unless a ZnTe nucleation layer is used [66].

Figure 3.17 presents the RHEED patterns and XRD rocking curves for CdTe buffer layers grown on GaSb substrates under varying VI/II flux ratios. It is observed that higher VI/II ratios, such as 2.5 and 3, lead to improved material quality, as demonstrated by narrower XRD FWHM values (71 arcsec) and long streaks in the RHEED patterns. These results highlight the importance of maintaining an adequate supply of Te atoms at the growth front to achieve high-quality CdTe buffer layers.

By further optimizing the growth temperature and VI/II flux ratio, CdTe buffer layers can be grown on GaSb with a material quality comparable to those grown on

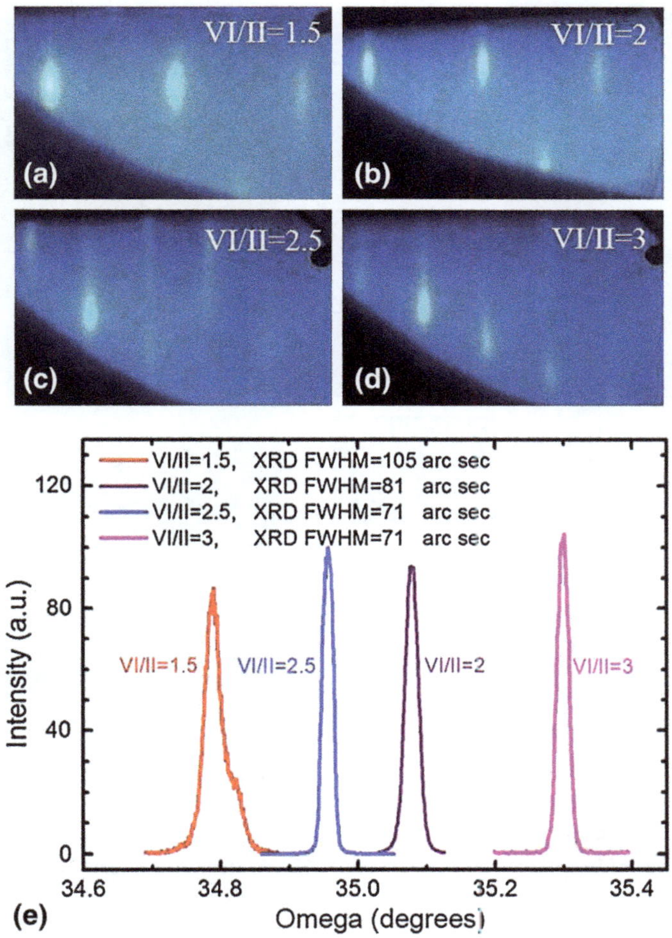

Figure 3.17. Representative RHEED patterns during growth of CdTe buffer layers under a VI/II flux ratio of (a) 1.5, (b) 2, (c) 2.5, and (d) 3, and (e) their corresponding XRD rocking curves. Reproduced from [7] with permission from Springer Nature.

GaAs. Figure 3.18 shows representative RHEED patterns, XRD rocking curves, and surface images (post-EPD etching) of CdTe buffer layers on GaSb (figures 3.18 (a) and (b)) and GaAs (figures 3.18(c) and (d)) substrates (all grown under optimized conditions). Clear, uniform, long, and narrow streaks in the RHEED patterns are observed for both types of substrates, indicating good crystallinity. The XRD FWHM values are 61 arcsec for CdTe on GaSb and 63 arcsec for CdTe on GaAs. Correspondingly, the EPD values are 5×10^6 cm^{-2} and 3×10^6 cm^{-2} for CdTe buffer layers on GaSb and GaAs, respectively. Note that the EPD values in CdTe buffer layers were measured after standard Everson EPD etching at UWA [68].

Table 3.2 lists the main statistical material characteristics of CdTe buffer layers grown on both GaSb and GaAs substrates at UWA. Generally, uniform, long, and narrow streak RHEED patterns indicate a smooth growth front and thus a high-quality

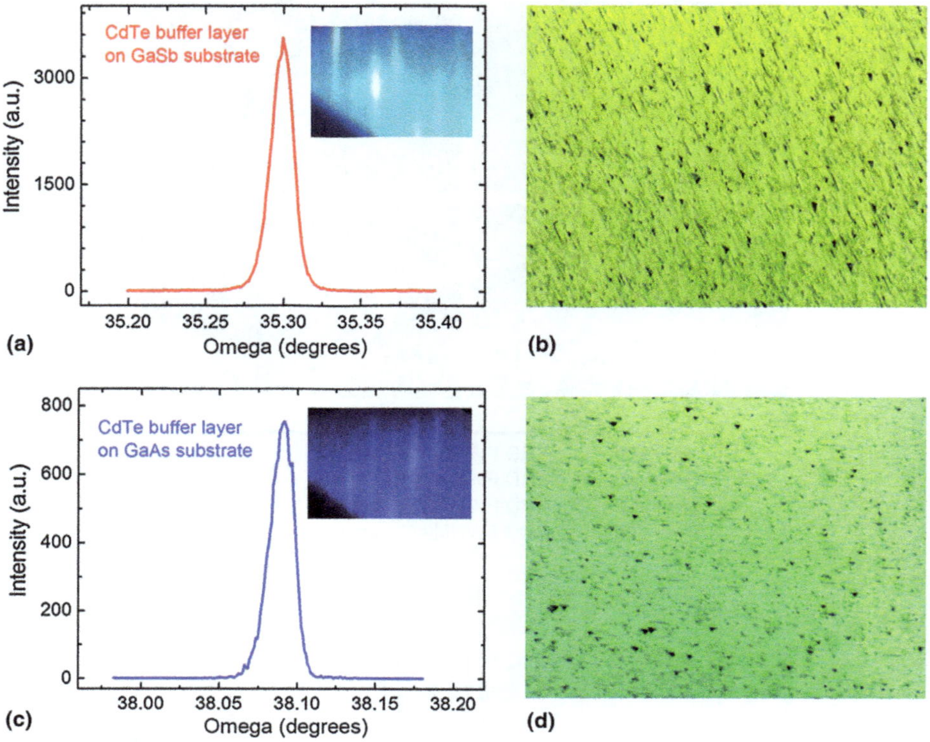

Figure 3.18. Representative XRD rocking curves and surface images (after EPD etching) of CdTe buffer layers on (a, b) GaSb and (c, d) GaAs substrates grown under optimized growth conditions. Insets in (a) and (c) show representative RHEED patterns during the growth of CdTe buffer layers on GaSb and GaAs, respectively. The size of images (b) and (d) is 95 μm × 65 μm. Reproduced from [7] with permission from Springer Nature.

Table 3.2. Main material characteristics of CdTe buffer layers grown at UWA on GaSb and GaAs substrates. Reproduced from [7] with permission from Springer Nature.

Alternative substrate	CdTe buffer layer thickness (μm)	XRD FWHM (arcsec)	RHEED pattern during CdTe buffer layer growth	Etch pit density ($\times 10^6$ cm^{-2})
GaAs	6–7	60–75	Long and uniform streaks	3–50
GaSb	3–7	55–71	Long and uniform streaks	5–30

MBE growth, while smaller XRD FWHMs indicate higher crystalline quality as defects can broaden the XRD FWHMs. As shown in table 3.2, statistically CdTe buffer layers grown on GaSb have a material quality comparable to and even slightly better than that on GaAs. It is noted that the main characteristics of CdTe buffer layers grown on GaAs at UWA is comparable to those of state-of-the-art CdTe buffer layer technology on GaAs reported recently [45, 66, 69].

As discussed in section 3.1, one of the main benefits for GaSb alternative substrates is the much smaller lattice mismatch between HgCdTe and GaSb, in comparison to other alternative substrates such as Si, Ge, and GaAs. So, it is expected that the dislocation density in CdTe and HgCdTe epilayers induced by the combined lattice/CTE mismatch will be much lower compared with that on other alternative substrates. To validate this, cross-sectional TEM measurements were undertaken at UWA to characterize the lattice mismatch/dislocation generation in the CdTe buffer layers [64]. Figure 3.19 presents the high-resolution TEM images and their related inverse fast Fourier transform (FFT) pattern near the CdTe/substrate interface [64]. It can be observed that the diffraction pattern of the CdTe layer overlaps that of the substrate, which indicates that they present the same face-centered cubic zinc blende structure for CdTe on both GaSb and GaAs substrates. The FFT technique was used to filter the images in figures 3.19(a) and (b) to highlight specific lattice fringes and misfit dislocations [64]. Figures 3.19(c) and (e) show the inverse FFT images of figure 3.19(a) by selecting the {111} and {−111} diffraction spots in the FFT pattern, while figures 3.19(d) and (f) show the inverse FFT images of figure 3.19(b) by selecting the {111} and {−111} diffraction spots in the FFT pattern. As indicated by the arrows in figures 3.19(c)–(f), misfit dislocations are observed on both the (111) and (−111) fringes. But, obviously the number of misfit dislocations near the CdTe/GaSb interface is observed to be much lower than that between CdTe and GaAs. Although TEM measurements only study a very small area of the samples, the results in figure 3.19 can still suggest that the misfit dislocation density in CdTe layers grown on GaSb is much lower than that on GaAs on average, which accords well with the main benefit of GaSb alternative substrates as proposed in section 3.1.

After achieving the direct CdTe buffer layers as discussed above, MWIR HgCdTe epilayers were grown on these CdTe/GaSb and CdTe/GaAs substrates using the typical growth parameters for growing HgCdTe at UWA ($T_g \sim 190$ °C, Hg flux $\sim 10^{-4}$ Torr, CdTe ~mid-10^{-7} Torr, and Te flux $\sim 10^{-6}$ Torr). The statistical material characteristics of HgCdTe epilayers grown on both CdTe/GaSb and CdTe/GaAs substrates are also summarized in table 3.3 [64]. Generally, HgCdTe layers grown on GaSb substrates present material quality comparable to those on GaAs substrates developed at UWA in terms of RHEED pattern during material growth, XRD FWHM, and EPD. Although the average EPDs in these direct CdTe buffer layers and HgCdTe layers grown on GaSb are comparable to those on GaAs, they still cannot meet the critical EPD level required for making high performance LWIR HgCdTe detectors. Therefore, further treatment of these direct CdTe buffer layers and HgCdTe layers is needed in order to improve the material quality to meet the requirements for fabricating high-performance MWIR and LWIR HgCdTe detectors.

3.5.2.2 Transitional buffer layer technology

Most recently, at UWA we proposed a unique transitional buffer layer technology to further enhance the material quality of CdTe buffer layers and HgCdTe layers grown on GaSb substrates [70]. Fundamentally, the misfit dislocations in the CdTe

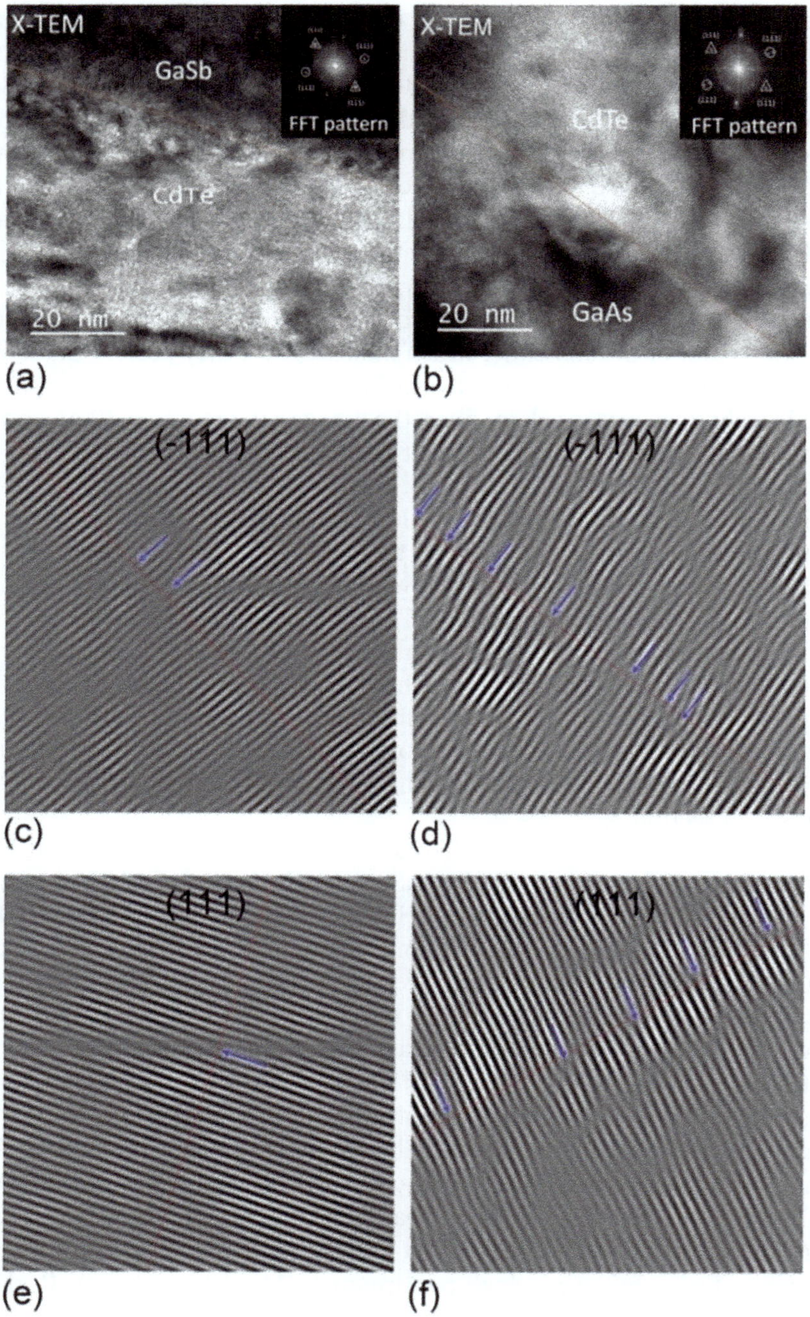

Figure 3.19. Representative high resolution TEM images of the (a) CdTe/GaSb (211)B interface, and (b) CdTe/GaAs (211)B interface, and their related inverse FFT images ((c) and (e) for CdTe/GaSb) and (d) and (f) for CdTe/GaAs) obtained by selecting {111} and {−111} diffraction spots. Insets of (a) and (b) show their related FFT pattern, in which {111} and {−111} diffraction spots are marked by triangles and circles, respectively. Reproduced from [64] with permission from Springer Nature.

Table 3.3. Relevant statistical material characteristics of HgCdTe layers MBE-grown at UWA on GaSb and GaAs substrates. Reproduced from [64] with permission from Springer Nature.

Alternative substrate	x value of $Hg_{1-x}Cd_xTe$ layer	HgCdTe layer thickness (μm)	CdTe buffer layer thickness (μm)	XRD FWHM (arcsec)	RHEED pattern during HgCdTe layer growth	Etch pit density ($\cdot 10^6$ cm^2)
GaSb	0.27–0.32	4.5–5.3	5.3–5.7	122–139	Long and uniform streaks	2–10
GaAs	0.26–0.32	5.7–6.7	5.7–6.7	98–155	Long and uniform streaks	8–40

buffer layers and HgCdTe layers grown on lattice mismatched substrates are caused by the large lattice mismatch and the consequent strain accumulation and relaxation. Therefore, the generation of misfit dislocations and thus threading dislocations can be effectively suppressed and reduced if the strain accumulated can be better accommodated/relaxed via appropriate approaches. The unique transitional buffer layer we proposed can effectively accommodate/relax the misfit strain in the system when growing CdTe and HgCdTe on GaSb substrates, and thus effectively suppress and reduce the generation of misfit dislocations and improve the material quality of CdTe buffer layers and HgCdTe layers. Compared with the direct CdTe buffer layer technology, this unique transitional buffer layer technology is to incorporate a unique thin Zn(Cd)Te-based transitional buffer layer before growing the thick CdTe buffer layer. As reported in our previous work [70], this transitional buffer layer consisted of approximately 30–50 periods of alternating layers of ZnTe/CdTe grown to a total thickness in the range of 150–200 nm. The main purposes of this thin transitional buffer layer are (1) to reduce the generation of misfit dislocations, (2) to prevent their propagation into the overlaying CdTe layer, and (3) to act as a defect and impurity gettering layer. These effects will significantly reduce the dislocation density in the CdTe buffer layer and HgCdTe layer grown subsequently.

Figure 3.20 shows the representative XRD rocking curves of CdTe buffer layers grown with and without incorporating a transitional buffer layer (TBL) between the CdTe buffer layer and the GaSb substrate [70]. For the convenience of the following discussion, the CdTe buffer layer grown with incorporating a TBL is labeled as 'Sample TBL,' while the CdTe buffer layer grown without incorporating a TBL is labeled as 'Sample OTBL.' In comparison to Sample OTBL, Sample TBL presents a smaller XRD FWHM. The XRD FWHM is 55.8 and 61.2 arcsec for Sample TBL and Sample OTBL, respectively. As discussed before, generally XRD FWHM provides an effective measure of material crystal quality as defects in the crystal structure can broaden the XRD peak. Therefore, the smaller XRD FWHM of Sample TBL suggests a higher crystal quality and a lower dislocation density in comparison to Sample OTBL.

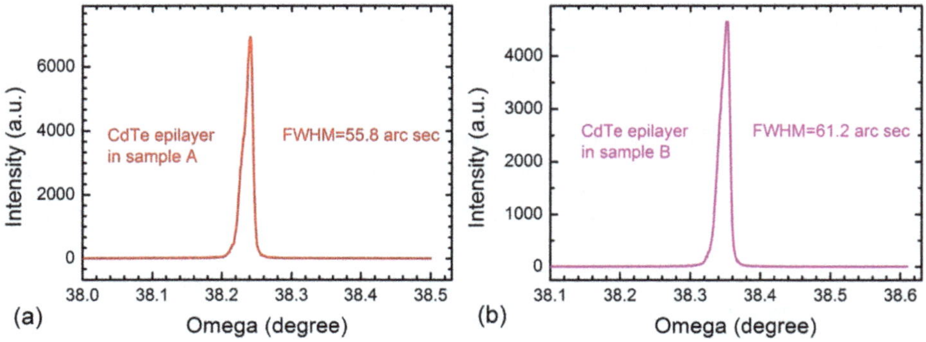

Figure 3.20. Representative XRD rocking curves of (a) Sample TBL and (b) Sample OTBL. Reprinted from [70], Copyright © 2018, with permission from Elsevier B.V. All rights reserved.

Table 3.4. Main material characteristics of CdTe buffer layers MBE-grown on GaSb substrates at UWA with and without incorporating a TBL between the CdTe buffer layer and the GaSb substrate.

Sample	XRD FWHM (arcsec)	RMS surface roughness (Å)	Etch pit density (x 10^5 cm^{-2})	Strain measured with XRD RSM
Sample TBL	55.8	9.8	1.4	2.25×10^{-6}
Sample OTBL	61.2	11.8	48	1.05×10^{-5}

Table 3.4 lists the main material characteristics of CdTe buffer layers grown at UWA on GaSb substrates with and without incorporating a TBL between the CdTe buffer layer and the GaSb substrate [70]. As shown in table 3.4, Sample TBL presents a much smoother surface compared to Sample OTBL. Note that the RMS surface roughness was measured with AFM. Statistically, the RMS surface roughness is 9.8 and 11.8 Å for Sample TBL and Sample OTBL, respectively. A surface roughness of 9.8 and 11.8 Å indicates a very high-quality layer-by-layer MBE growth for the CdTe epitaxial layer considering that the starting GaSb substrate has a surface roughness of 4.9 Å as measured by AFM [70]. Compared with Sample OTBL, the smaller RMS surface roughness of Sample TBL indicates that the lattice mismatch and thus misfit strain between CdTe and GaSb has been better accommodated in Sample TBL, which facilitates smoothing out the MBE growth front and results in a smoother surface.

As listed in table 3.4, statistically Sample TBL shows an EPD much lower than Sample OTBL: an average EPD of 1.4×10^5 cm^{-2} for Sample TBL, while 4.8×10^6 cm^{-2} for Sample OTBL. This low EPD value for CdTe buffer layers grown incorporating a TBL is mainly caused by the better relaxed misfit strain and thus lower generation of misfit dislocations in the system [70]. In principle, because the CdTe layer in Sample OTBL is significantly thicker than that in Sample TBL, the EPD for Sample OTBL would be expected to be lower than that for Sample TBL if the generation of misfit dislocations at the substrate/epilayer interface was the same for both samples. Therefore, the much lower EPD in Sample TBL can only be due to the beneficial effects of the transitional buffer layer since otherwise, the two samples are identical. The main benefits due to the transitional buffer layer could include (1) improved relaxation of the lattice mismatch close to the GaSb/CdTe interface, (2) filtering/blocking the propagation of any interface-generated misfit dislocations, and/or (3) gettering of defects and impurities from the subsequently grown CdTe layer.

As discussed above, one of the main benefits for the transitional buffer layer is to better accommodate/relax the misfit strain in the system, and thus effectively suppress and reduce the generation of misfit dislocations and threading dislocations. To better understand the relaxation of lattice mismatch and misfit strain in the grown CdTe layers, high-resolution XRD reciprocal space mapping (RSM) measurements were undertaken on both Sample TBL and Sample OTBL. Generally, in-plane strain caused by lattice mismatch will result in a shift of the XRD peak of the epitaxial layer relative to that of the substrate. In this UWA study, the XRD reciprocal space maps were measured with a symmetrical (422) scan which can provide accurate information on the in-plane strain caused by the lattice mismatch/strain relaxation in the CdTe layers. By following the strain analysis procedure reported previously [71, 72], the strain in Sample TBL and Sample OTBL can be determined, the results of which are listed in table 3.4. It is observed that the strain in Sample TBL (2.25×10^{-6}) is much lower than that in Sample OBTL (1.05×10^{-5}), and even approaches the magnitude of -0.62×10^{-6} for HgCdTe layers grown on lattice matched CdZnTe substrates [73]. This suggests that the CdTe/GaSb lattice mismatch in Sample TBL is much better relaxed than that in Sample OTBL. This better relaxed misfit strain should result in less generation of misfit dislocations, which correlates well with the lower EPD levels observed for Sample TBL.

This result is very encouraging as this is a recent effort, and the average EPD of CdTe layers after incorporating a TBL is already below the critical EPD level (5×10^5 cm^{-2}) required for fabricating LWIR HgCdTe detectors, and is close to the EPD (mid-10^4 cm^{-2} to low-10^5 cm^{-2}) in commercial CdZnTe substrates. An even lower EPD ($\leq\sim 10^5$ cm^{-2}) can be expected for CdTe epitaxial layers by optimizing the TBL, growing a thicker CdTe layer (≥ 10 μm) [10, 12, 36, 42] and implementing other treatments such as cyclic thermal annealing [13, 52] to further reduce the EPD in the CdTe buffer layers.

3.5.2.3 Strained superlattice dislocation filtering layers

As discussed above, transitional buffer layer technology has led to an EPD in the CdTe layer much lower than those reported on Si, Ge, and GaAs substrates. However, the EPD achieved is still higher than that of the state-of-the-art CdZnTe substrates. Generally, the dislocations in CdTe buffer layer can be reduced via two ways: (1) to suppress the dislocation generation via better strain relaxation. The transitional buffer layer technology better relaxes the strain and thus reduces/supresses the dislocation generation in the CdTe buffer layers; (2) to control the propogation direction of threading dislocations and thus annihilate the threading dislocations within. Theoretically, strained-layer superlattices can be used to bend and control the dislocation propagation direction towards the growth plane, and thus reduce the ultimate dislocation density in the top epitaxial layers. This concept has been demonstrated in the growth of III–V semiconductors on lattice mismatched substrates [10–12], and the dislocation density was reduced from $\sim 10^9$ cm^{-2} to $\sim 10^5$ cm^{-2} in GaAs buffer layers grown on Si by incorporating InGaAs/GaAs strained-layer superlattices, leading to high-performance InAs/GaAs quantum dot lasers. The CdZnTe/CdTe superlattice DFLs have also been utilized to reduce the dislocation density in CdTe buffer layers grown on GaAs as discussed in section 3.4. Similarly, strained CdZnTe/CdTe superlattice layers can also be used as DFLs for reducing the threading dislocations and thus improving the material quality of CdTe buffers grown on GaSb (211)B substrates. Considering the smaller lattice mismatch between CdTe and GaSb (6.1%) in comparision to that between CdTe and GaAs (14%), the CdZnTe/CdTe superlattice layers should have a better dislocation reduction effect for CdTe on GaSb.

The UWA research group has undertaken comprehensive study on CdZnTe/CdTe superlattice DFLs for dislocation reduction in CdTe buffer layers grown on GaSb substrates. The reported MBE growth process for CdTe/GaSb buffer layer with incorporating CdZnTe/CdTe superlattice DFLs is as follows [74]:

CdTe buffer layers were grown on GaSb (211)B substrates using a Riber 32P MBE system equipped with effusion cells of Zn, CdTe, and Te. Figure 3.21 shows the schematic sample structures for growing CdTe on GaSb with and without CdZnTe/CdTe superlattice layer DFLs. The substrate temperature was first ramped up to 520 °C for 2 min without the protection of background Sb flux to remove the native oxide on the substrate surface. After the oxide desorption, the substrates were then cooled down to 280 °C, and then a 1850 nm CdTe layer was grown. After this,

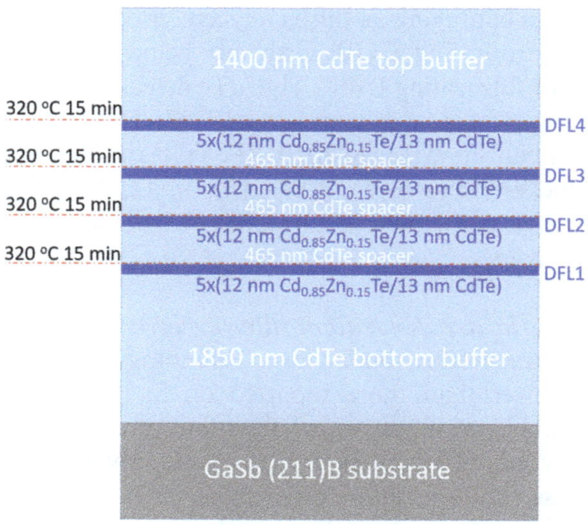

Figure 3.21. Schematic sample structure for the MBE growth of CdTe buffer on GaSb (211)B with strained CdZnTe/CdTe strained superlattice layers. Reproduced from [74] with permission from Springer Nature.

four sets of CdZnTe/CdTe DFLs separated by a 470 nm CdTe spacer layer each were grown at 285 °C and 280 °C, respectively. Each set of $Cd_{0.85}Zn_{0.15}Te$/CdTe DFL consisted of five thin layers of $Cd_{0.85}Zn_{0.15}Te$ (12 nm) separated by four thin layers of CdTe (13 nm). It is worth noting that in each set of DFLs the thickness of $Cd_{0.85}Zn_{0.15}Te$ layers, 0.9% lattice mismatch to CdTe, was designed to be below its theoretical critical thickness of 28 nm [75] in order to avoid lattice relaxation and thus generation of new misfit dislocation and threading dislocations. *In-situ* thermal annealing of the CdZnTe/CdTe superlattice layers was also undertaken four times in order to enhance dislocation annihilation [76], which was achieved by raising the substrate temperature to 320 °C for 15 min. The sample growth was finished by depositing a 1400 nm CdTe layer at 280 °C. Note that the sample grown with these procedures has a total thickness of 5150 nm. For comparision, a reference sample of a CdTe buffer layer with a thickness of 4900 nm was also grown at 280 °C with the same CdTe and Te fluxes as those of the DFL sample, but without any CdZnTe/CdTe superlattice layers. *In-situ* RHEED was used to monitor the growth front of the sample surface during the sample growth.

Apart from XRD measurements, RHEED and AFM were also utlized to understand and evaluate the MBE growth and material qualtiy of CdTe buffer layers with and without the incorporation of CdZnTe/CdTe superlattice DFLs. As evident from figures 3.22(a) and (d), the spotty RHEED patterns during the MBE growth of CdTe become streaky with the incorporation of CdZnTe/CdTe DFLs, indicating an improvement in material quality. Correspondingly, as shown by the AFM images in figures 3.22(b) and (e) the pitted surface defects which likely originated from the lattice mismatch and non-ideal oxide desorption of GaSb are

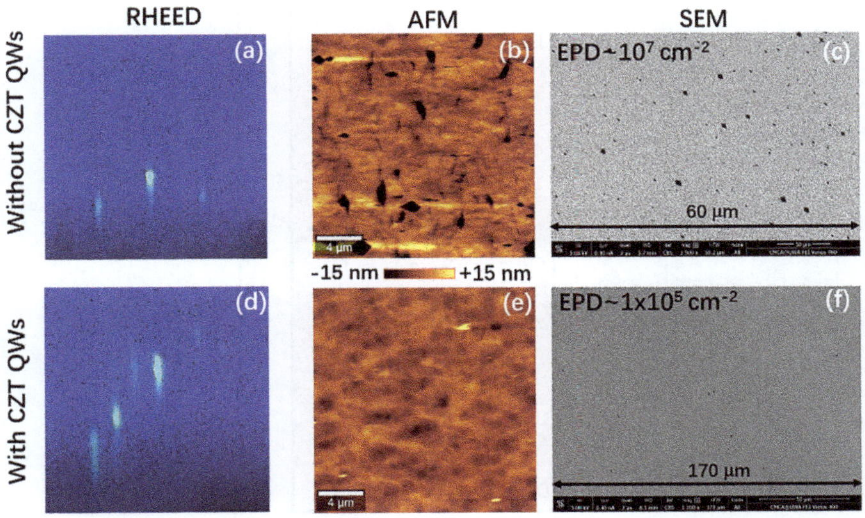

Figure 3.22. RHEED patterns during MBE growth, AFM images, SEM images for EPD measurements of the uppermost CdTe surface. Panels (a), (b) and (c) are for a 4900-nm-thick single CdTe layer, and (d), (e) and (f) are for a 5145-nm-thick CdTe buffer layer that includes the CdZnTe/CdTe DFLs. Reproduced from [74] with permission from Springer Nature.

significantly reduced due to the presence of DFLs. The surface roughness of the CdTe with DFLs is measured to be 1.7 nm which is lower than the 3.8 nm measured on the CdTe-only reference sample. To evaluate the effect of the DFLs on the threading dislocation density in the CdTe layer grown, EPD measurements were undertaken by using a 60 s Everson etch on both types of samples, corresponding to a CdTe etch depth of less than 1.4 μm. From the SEM images in figures 3.22(c) and (f), the EPD for the uppermost CdTe with DFLs is measured to be $\sim 1 \times 10^5$ cm^{-2}, which is approximately two orders of magnitude lower than that of the single-layer CdTe reference sample, with an EPD of $\sim 2 \times 10^7$ cm^{-2}. This result can be attributed to the dislocation filtering effect of the strained CdZnTe/CdTe superlattice layers.

In order to assess the filtering efficiency of individual CdZnTe/CdTe DFL, the sample with multiple DFLs was etched down to different depths, and SEM imaging was undertaken to determine the EPD values within the individual CdTe layers. First, the sample was etched to a depth of approximately 2.6 μm using a 2 min etching time within a photoresist-free area. The etched surface thus corresponds to the mid-region of the CdTe layer (near the upper interface of the DFL2/CdTe) and is labeled 'B' in the schematic sample structure shown in figure 3.23(a). Figure 3.23(c) shows a representative magnified SEM image of surface 'B,' indicating that there are three main types of etch pits: (1) conventional triangular etch pits that originate from

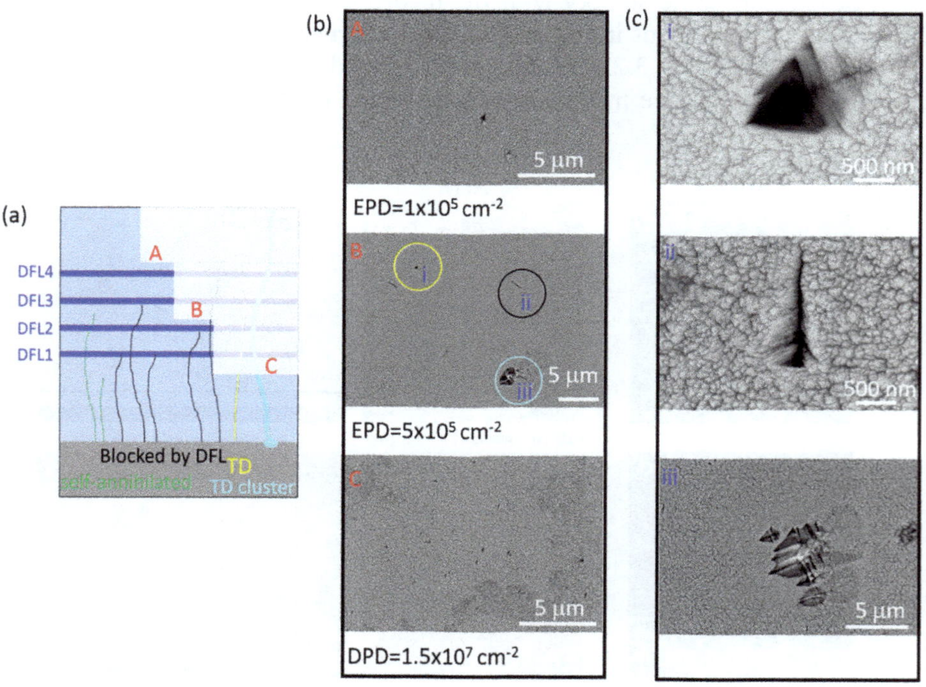

Figure 3.23. (a) Schematic diagram of the CdTe buffer layer including four DFL structures used for EPD measurements at different etch depths within layers A, B, and C. (b) Representative SEM images of surfaces A, B, and C after EPD etching. (c) Magnified images of three different types of etch pits observed in the image 'B' of (b). Reproduced from [74] with permission from Springer Nature.

threading dislocations that penetrate through the DFL region; (2) quasi-etch pits that contribute the majority of pits, and are expected to have originated from threading dislocations that have been diverted by the DFL layer and/or due to self-annihilation; and (3) a small number of large area (3–5 μm diameter) dislocation clusters, which are likely to be caused by the non-ideal thermal desorption of oxides from the GaSb substrate surface prior to MBE growth, and penetrate through the entire buffer layer since they can be observed by optical microscopy before EPD etching. The total EPD is determined to be $\sim 5 \times 10^5$ cm^{-2} for surface 'B,' which is calculated by averaging the pit counts over five large-area SEM images.

After evaluating the EPD of surface 'B' within the middle CdTe layer, the photoresist used for the first etch was removed, and both the unetched area and etched area were etched for a period of 60 s. This leads to the exposure of two etched surfaces with depths of 1.4 and 4.0 μm, which correspond to positions in the uppermost CdTe layer above the DFL region and a position in the bottom CdTe layer below the DFL region, and are labelled 'A' and 'C' in figure 3.23(a), respectively. Figure 3.23(b) presents the representative SEM images of surfaces 'A' and 'C' respectively, and similar types of etch pits are evident. The EPD of CdTe surface 'A' (above the DFL region) is determined to be $\sim 1 \times 10^5$ cm^{-2}, while that of surface 'C' (below the DFL region) is determined to be as high as 1.5×10^7 cm^{-2}. In principle, the density of threading dislocations should reduce somewhat with increasing thickness of a single-layer CdTe buffer due to self-annihilation. However the efficiency of such a self-annihilation process is relatively low, and no significant EPD reduction with thickness is observed in the reference CdTe sample without incorporating DFLs. This suggests that the DFLs are effective filters for reducing the propagation of threading dislocations to the uppermost CdTe layer.

As indicated in figure 3.24, the changes in EPD values by incorporating different sets of DFLs can be experimentally fitted by a power relationship: $\rho = \rho_0(1 - \eta)^N$,

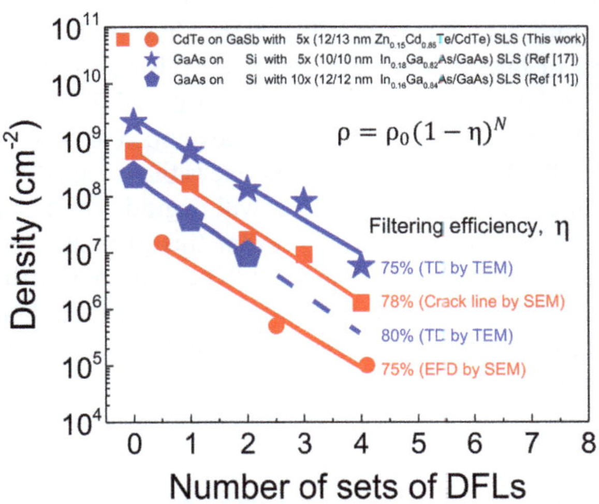

Figure 3.24. Dependence of defect density (EPD) on the number of sets of DFLs. Reproduced from [74] with permission from Springer Nature.

where ρ is the EPD value of the CdTe layer after incorporating a certain number of DFLs, ρ_0 is the initial EPD of the bottom CdTe layer, N is the number of DFLs, and η is the filtering efficiency. The filtering efficiency η of each DFL can be determined to be ~70%. These results are very promising, since they suggest that even lower EPD values can be achieved by further optimizing the structural parameters of the DFL to increase the filtering efficiency, improving the structural parameters of the DFL and the growth conditions of the epitaxial layers through better thermal cleaning of the GaSb substrate surface, and/or by increasing the number of DFLs within the buffer layer.

In all, GaSb presents a promising alternative substrate for MBE growth of high-quality CdTe and HgCdTe layers. With the unique transitional buffer layer technology and strained superlattice DFL technology, the dislocation density in a CdTe buffer layer can be well reduced and controlled with the lowest EPD level of 1.0×10^5 cm^{-2}, which is below the critical EPD level of 5×10^5 cm^{-2}, and lower than the EPD values reported on other alternative substrates including Si, Ge, and GaAs. Although HdCdTe layers haven't been grown on these CdTe/GaSb wafers, high quality HgCdTe layers are expected as CdTe buffer layer will act as a template for growing the subsequent HgCdTe layers, and determine the minimum dislocation level to be achieved. These early results are very promising, especially given the short timeframe of this development. With continued refinement of the growth parameters and the oxide desorption process, even higher quality CdTe layers on GaSb are expected. Overall, these findings strongly suggest the potential of GaSb as a next-generation alternative substrate for MBE growth of HgCdTe IR materials.

3.6 Other alternative substrates

As discussed before, Si, Ge, GaAs, and GaSb alternative substrates have been widely studied over the past several decades. Apart from those, InSb substrate also provides a potential alternative substrate for growing high-quality HgCdTe, especially LWIR and VLWIR HgCdTe. Note that InSb and LWIR HgCdTe have almost identical lattice constants and CTE. However, there were only very limited reports on MBE growth of HgCdTe on InSb [77]. There are a number of challenges to be addressed before InSb can be developed as a substrate material for the growth of HgCdTe. The most challenging one is the strong diffusion of In and Sb atoms into CdTe and HgCdTe layers which will degrade their electrical properties. Diffusion-blocking buffer layers do not prevent this problem, and removal of the substrate before post-growth annealing would be required with the use of InSb substrates as shown in the work by Jaime-Vasquez *et al* [78]. This complicates the development of InSb alternative substrate technology for growing high-quality HgCdTe IR materials.

3.7 Summary

With intensive study over the past several decades, significant progress has been made on alternative substrates for growing high-quality CdTe and HgCdTe layers. Table 3.5 summarizes the material quality of CdTe and HgCdTe layers grown on

Table 3.5. Typical XRD FWHM and EPD values for CdTe and/or HgCdTe epilayers grown on CdZnTe, Si, Ge GaAs, and GaSb substrates reported in references [12, 36, 42, 70].

As-grown (211) B	CdZnTe substrate	HgCdTe /CdZnTe	CdTe /Si	HgCdTe/ Si	HgCdTe/ Ge	CdTe /GaAs	HgCdTe/ GaAs	CdTe /GaSb
XRD FWHM (arcsec)	8	12	50	80	77	30	80	56
EPD (cm^{-2})	5×10^4	3×10^4	5×10^6	4×10^6	5×10^6	2×10^6	2×10^6	1×10^5

CdZnTe, Si, Ge, GaAs, and GaSb substrates [12, 36, 42, 70]. Obviously, there is still a substantial gap between the material quality of CdTe and HgCdTe grown on alternative substrates compared to those grown on CdZnTe, especially LWIR HgCdTe materials. Despite this, Si, Ge, and GaAs alternative substrates have been used for producing SWIR and MWIR HgCdTe detectors and their FPAs with performance comparable to those grown on CdZnTe substrates. However, the performance of LWIR HgCdTe detectors and their FPAs grown on alternative substrates is still much lower than those on CdZnTe. Significant effort is still needed to further reduce the dislocation density in the CdTe and HgCdTe layers, and enhance the HgCdTe material quality and thus their IR detector device performance to achieve their ultimate industry applications.

References

[1] Lawson W D, Nielson S, Putley E H and Young A S 1959 Preparation and properties of HgTe and mixed crystals of HgTe-CdTe *J. Phys. Chem. Solids* **9** 325

[2] https://teledyneimaging.com/en/aerospace-and-defense/products/sensors-overview/infrared-hgcdte-mct/hawaii-4rg/

[3] Rogalski A 2005 HgCdTe infrared detector material: history, status and outlook *Rep. Prog. Phys.* **68** 2267

[4] Norton P 2002 HgCdTe infrared detectors *Opto-Electron. Rev.* **10** 159

[5] Rogalski A, Antoszewski J and Faraone L 2009 Third-generation infrared photodetector arrays *J. Appl. Phys.* **105** 091101

[6] Garland J W and Sivananthan S 2010 Chapter 32 - Molecular-beam epitaxial growth of HgCdTe *Springer Handbook of Crystal Growth* ed G Dhanaraj, K Byrappa, V Prasad and M Dudley (Berlin Heidelberg: Springer) pp 1084–7

[7] Lei W, Gu R J, Antoszewski J, Dell J and Faraone L 2014 GaSb: a new alternative substrate for epitaxial growth of HgCdTe *J. Electron. Mater.* **43** 2788

[8] Jin Kim J 2012 Characterization of HgCdTe and related materials and substrates for third generation infrared detectors *PhD thesis* (Arizona State University)

[9] Kim J, Jacobs R N, Almeida L A, Jaime-Vasquez M, Nozaki C and Smith D J 2013 TEM characterization of HgCdTe/CdTe grown on GaAs (211) B substrates *J. Electron. Mater.* **42** 3142

[10] Carmody M *et al* 2012 Recent progress in MBE growth of CdTe and HgCdTe on (211)B GaAs substrates *J. Electron. Mater.* **41** 2719

[11] He L *et al* 2007 MBE HgCdTe on Si and GaAs substrates *J. Cryst. Growth* **301–302** 268
[12] Benson J D *et al* 2012 Growth and analysis of HgCdTe on alternate substrates *J. Electron. Mater.* **41** 2971
[13] Lei W, Antoszewski J and Faraone L 2015 Progress, challenges and opportunities for HgCdTe infrared materials and detectors *Appl. Phys. Rev.* **2** 041303
[14] Jówikowski K and Rogalski A 2000 Effect of dislocations on performance of LWIR HgCdTe photodiodes *J. Electron. Mater.* **29** 736–41
[15] Johnson S M, Rhiger D R, Rosbeck J P, Peterson J M, Taylor S M and Boyd M E 1992 Effect of dislocations on the electrical and optical properties of long-wavelength infrared HgCdTe photovoltaic detectors *J. Vac. Sci. Technol.* B **10** 1499–506
[16] Vilela M F, Buell A A, Newton M D, Venzor G M, Childs A C, Peterson J M, Franklin J J, Bornfreund R E, Radford W A and Johnson S M 2005 Growth and control of middle wave infrared (MWIR) $Hg_{1-x}Cd_xTe$ on Si by molecular beam epitaxy *J. Electron. Mater.* **34** 898–904
[17] Carmody M *et al* 2005 Molecular beam epitaxy grown long wavelength infrared HgCdTe on Si detector performance *J. Electron. Mater.* **34** 832–8
[18] de Lyon T J *et al* 1998 Molecular beam epitaxial growth of HgCdTe infrared focal-plane arrays on Si substrates for mid-wave infrared applications *J. Electron. Mater.* **27** 550–5
[19] de Lyon T J, Jensen J E, Kasai I, Venzor G M, Kosai K, de Bruin J B and Ahlgren W L 2002 Molecular beam epitaxial growth and high-temperature performance of HgCdTe mid-wave infrared detectors *J. Electron. Mater.* **31** 220–6
[20] Johnson S M *et al* 2004 HgCdTe/Si materials for long wavelength infrared detectors *J. Electron. Mater.* **33** 526–30
[21] Reddy M *et al* 2011 Molecular beam epitaxy growth of HgCdTe on large-area Si and CdZnTe substrates *J. Electron. Mater.* **40** 1706
[22] Velicu S, Lee T S, Grein C H, Boieriu P, Chen Y P, Dhar N K, Dinan J and Lianos D 2005 Monolithically integrated HgCdTe focal plane arrays *J. Electron. Mater.* **34** 820–31
[23] Sporken R, Sivananthan S, Mohavadi K K, Monfroy G, Boukerche M and Faurie J P 1989 Molecular beam epitaxial growth of CdTe and HgCdTe on Si(100) *Appl. Phys. Lett.* **55** 1879–81
[24] Chen Y P, Brill G and Dhar N K 2003 MBE growth of CdSeTe/Si composite substrate for long-wavelength IR HgCdTe applications *J. Cryst. Growth* **252** 270–4
[25] Dhar N K, Wood C E C, Gray A, Wei H Y, Salamanca-Riba L and Dinan J H 1996 Heteroepitaxy of CdTe on {211} Si using crystallized amorphous ZnTe templates *J. Vac. Sci. Technol.* B **14** 2366
[26] Tsen G K O 2010 Investigation of molecular beam epitaxy grown p-type mercury cadmium telluride for infrared detector applications *PhD Thesis* University of Western Australia p 59
[27] Chen Y, Farrell S, Brill G, Wijewarnasuriya P and Dhar N 2008 Dislocation reduction in CdTe/Si by molecular beam epitaxy through *in-situ* annealing *J. Cryst. Growth* **310** 5303
[28] Speck J S, Brewer M A, Beltz G, Romanov A E and Pompe W 1996 Scaling laws for the reduction of threading dislocation densities in homogeneous buffer layers *J. Appl. Phys.* **80** 3808
[29] Almeida L A, Hirsch L, Martinka M, Boyd P R and Dinan J H 2001 Improved morphology and crystalline quality of MBE CdZnTe/Si *J. Electron. Mater.* **30** 608–10
[30] Chen Y P, Brill G and Dhar N K 2004 MBE growth of CdSeTe/Si composite substrate for long-wavelength IR HgCdTe applications *J. Cryst. Growth* **252** 270–4
[31] Chen Y P, Brill G, Campo E M, Hierl T, Hwang J C M and Dhar N K 2004 Molecular beam epitaxial growth of $Cd_{1-y}Zn_ySe_x\ Te_{1-x}$ on Si (211) *J. Electron. Mater.* **33** 498–502

[32] Brill G, Farrell S, Chen Y P, Wijewarnasuriya P S, Rao M V, Benson J D and Dhar N 2010 Dislocation reduction of HgCdTe/Si through ex situ annealing *J. Electron. Mater.* **39** 967

[33] Farrell S, Brill G, Chen Y, Wijewarnasuriya P S, Rao M V, Dhar N and Harris K 2010 Ex situ thermal cycle annealing of molecular beam epitaxy grown HgCdTe/Si layers *J. Electron. Mater.* **39** 43

[34] Boieriu P, Grein C H, Velicu S, Garland J, Fulk C, Stoltz A, Mason W, Bubulac L, DeWames R and Dinan J H 2006 Effects of hydrogen on majority carrier transport and minority carrier lifetimes in LWIR HgCdTe on Si *J. Electron. Mater.* **35** 1385

[35] Hu W D, Chen X S, Ye Z H and Lu W 2011 A hybrid surface passivation on HgCdTe long wave infrared detector with *in-situ* CdTe deposition and high-density hydrogen plasma modification *Appl. Phys. Lett.* **99** 091101

[36] He L *et al* 2007 MBE HgCdTe on Si and GaAs substrates *J. Cryst. Growth* **301–302** 268

[37] Garland J and Sporken R 2011 Substrates for the epitaxial growth of MCT *Mercury Cadmium Telluride Growth, Properties and Applications* ed P Capper and J Garland (New York: Wiley)

[38] Stoltz A J, Benson J D, Carmody M, Farrell S, Wijewarnasuriya P S, Brill G, Jacobs R and Chen Y 2011 Reduction of dislocation density in HgCdTe on Si by producing highly reticulated structures *J. Electron. Mater.* **40** 1785–9

[39] Zanatta J P, Ferret P, Duvaut P, Isselin S, Theret G, Rolland G and Million A 1998 Heteroepitaxy of CdTe on Ge(211) substrates by molecular beam epitaxy *J. Cryst. Growth* **184/185** 1297–301

[40] Zanatta J P, Duvaut P, Ferret P, Million A, Destefanis G, Rambaud P and Vannuffel C 1997 Growth of HgCdTe and CdTe (331)B on germanium substrate by molecular beam epitaxy *Appl. Phys. Lett.* **71** 2984

[41] Badano G *et al* 2005 High-resolution FPAs on MBE-grown HgCdTe/CdTe/Ge *Proc. SPIE* **5964** 596406

[42] Zanatta J P *et al* 2006 Molecular beam epitaxy growth of HgCdTe on Ge for third-generation infrared detectors *J. Electron. Mater.* **35** 1231

[43] Zanatta J P, Luchier N, Audebert P, Demars P, Chamonal J P, Ravetto M and Wolny M 1998 256 x 256 HgCdTe MWIR array grown on Ge substrates *Proc. of the SPIE* **3379** 586

[44] Vilela M F *et al* 2008 LWIR HgCdTe detectors grown on Ge substrates *J. Electron. Mater.* **37** 1465

[45] Jacobs R N *et al* 2012 Development of MBE II–VI epilayers on GaAs(211)B *J. Electron. Mater.* **41** 2707–13

[46] Arias J M, Zandian M, Shin S H, McLevige W V, Pasko J G and DeWames R E 1991 Dislocation density reduction by thermal annealing of HgCdTe epilayers grown by molecular beam epitaxy on GaAs substrates *AIP Conf. Proc.* **235** 1646–50

[47] Tyagi S *et al* 2024 Annealing of MBE-grown CdTe epitaxial layer at various tellurium overpressure for reduced defect density *J. Mater. Sci., Mater. Electron.* **35** 982

[48] Sasmaz E, Kaldirim M, Eker S, Tolungüc A and Uközer S 2019 Optimization of growth parameters for molecular beam epitaxial growth of (211)B CdTe layers on GaAs substrates *J. Electron. Mater.* **48** 6069

[49] Nishino H, Saito T and Nishijima Y 1996 VI/II ratio dependence of surface macrodefects in CdTe/ZnTe/GaAs(100) growth by metalorganic vapor phase epitaxy *J. Cryst. Growth* **165** 227–32

[50] Sasaki T, Tomono M and Oda N 1995 Crystallinity improvement of HgCdTe on GaAs grown by molecular beam epitaxy *J. Cryst. Growth* **150** 785–9
[51] Sasaki T and Oda N 1995 Dislocation reduction in HgCdTe on GaAs by thermal annealing *J. Appl. Phys.* **78** 3121–4
[52] Shin S H, Arias J M, Edwall D D, Zandian M, Pasko J G and DeWames R E 1992 Dislocation reduction in HgCdTe on GaAs and Si *J. Vac. Sci. Technol.* B **10** 1492–8
[53] Bakali E, Selamet Y and Tarhan E 2018 Effect of annealing on the density of defects in epitaxial CdTe (211)/GaAs *J. Electron. Mater.* **47** 4780
[54] Sugiyama I, Hobbs A, Saito T, Ueda 0, Shinohara K and Takigawa H 1992 Dislocation reduction in HgCdTe epilayers on GaAs by using CdTe/CdZnTe strained-layer superlattices in CdTe layers *J. Cryst. Growth* **117** 161–5
[55] Pan W W, Gu R J, Zhang Z K, Lei W, Umana-Membreno G A, Smith D J, Antoszewski J and Faraone L 2022 Defect engineering in MBE-grown CdTe buffer layers on GaAs (211) B substrates *J. Electron. Mater.* **51** 4869–83
[56] Pan W W, Faraone L and Lei W 2024 Dislocation filtering layers for defect reduction in the heteroepitaxial growth of infrared optoelectronic materials *CdTe and CdZnTe Materials: Material Properties and Applications* ed K Iniewski (Switzerland: Springer Nature) pp 23–6
[57] Million A, Dhar N K and Dinan J H 1996 Heteroepitaxy of CdTe on {211} Si substrates by molecular beam epitaxy *J. Cryst. Growth* **159** 76–80
[58] Ward T, Sánchez A M, Tang M, Wu J, Liu H, Dunstan D J and Beanland R 2014 Design rules for dislocation filters *J. Appl. Phys.* **116** 063508
[59] MacPherson G and Goodhew P J 1996 A refined scheme for the reduction of threading dislocation densities in $In_xGa_{1-x}As$/GaAs epitaxial layers *J. Appl. Phys.* **80** 6706–10
[60] Li W, Chen S, Tang M, Wu J, Hogg R, Seeds A, Liu H and Ross I 2018 Effect of rapid thermal annealing on threading dislocation density in III–V epilayers monolithically grown on silicon *J. Appl. Phys.* **123** 215303
[61] Miles R H, McGill T C, Chow P P, Johnson D C, Hauenstein R J, Nieh C W and Strathman M D 1988 Dependence of critical thickness on growth temperature in $Ge_x\ Si_{1-x}$/Si superlattices *Appl. Phys. Lett.* **52** 916–8
[62] Jain S C and Hayes W 1991 Structure, properties and applications of Ge_xSi_{1-x} strained layers and superlattices *Semicond. Sci. Technol.* **6** 547
[63] Zou J, Cockayne D J H and Usher B F 1993 Misfit dislocations and critical thickness in InGaAs/GaAs heterostructure systems *J. Appl. Phys.* **73** 619–26
[64] Lei W, Gu R J, Antoszewski J, Dell J, Neusser G, Sieger M, Mizaikoff B and Faraone L 2015 MBE growth of mid-wave infrared HgCdTe layers on GaSb alternative substrates *J. Electron. Mater.* **44** 3180–7
[65] Lei W 2018 A review on the development of GaSb alternative substrates for the epitaxial growth of HgCdTe *J. Nanosci. Nanote. chnol.* **18** 7349
[66] He L *et al* 2008 MBE HgCdTe on alternative substrates for FPA applications *J. Electron. Mater.* **37** 1189
[67] Weiss E, Klin O, Grossman S, Greenberg S, Klipstein P C, Akhvlediani R, Tessler R, Edrei R and Hoffman A 2007 Hydrogen and thermal deoxidations of InSb and GaSb substrates for molecular beam epitaxial growth *J. Vacuum Sci. Technol.* A **25** 736–45
[68] Everson W J, Ard C K, Sepich J L, Dean B E, Neugebauer G T and Schaake H F 1995 Etch pit characterization of CdTe and CdZnTe substrates for use in mercury cadmium telluride epitaxy *J. Electron. Mater.* **24** 505–10

[69] Wenisch J, Eich D, Lutz H, Schallenberg T, Wollrab R and Ziegler J 2012 MBE growth of MCT on GaAs substrates at AIM *J. Electron. Mater.* **41** 2828–32
[70] Lei W, Ren Y L, Madni I and Faraone L 2018 Low dislocation density MBE process for CdTe-on-GaSb as an alternative substrate for HgCdTe growth *Infrared Phys. Technol.* **92** 96
[71] van der Sluis P 1994 Determination of strain in epitaxial semiconductor structures by high-resolution x-ray diffraction *Appl. Phys.* A **58** 129–34
[72] Sewell R H, Musca C A, Dell J M, Faraone L, Dieing T and Usher B 2004 Reciprocal space mapping of MBE-grown HgCdTe heterostructures *Proceeding of 2004 Conf. on Optoelectronic and Microelectronic Materials and Devices (Brisbane, Australia)* pp 81–4
[73] Madni I, Umana-Membreno G A, Lei W, Gu R J, Antoszewski J and Faraone L 2017 X-ray reciprocal space mapping of MBE grown HgCdTe on alternative substrates *Cryst. Res. Technol.* **52** 1700167
[74] Pan W W, Gu R J, Zhang Z K, Liu J L, Lei W and Faraone L 2020 Strained CdZnTe/CdTe superlattices as threading dislocation filters in lattice mismatched MBE growth of CdTe on GaSb *J. Electron. Mater.* **49** 6983
[75] Cibert J, André R, Deshayes C, Feuillet G, Jouneau P H, Dang L S, Mallard R, Nahmani A, Saminadayar K and Tatarenko S 1991 CdTe/ZnTe: critical thickness and coherent heterostructures *Superlattices Microstruct.* **9** 271–4
[76] George I, Becagli F, Liu H Y, Wu J, Tang M and Beanland R 2015 Dislocation filters in GaAs on Si *Semicond. Sci. Technol.* **30** 114004
[77] Pan W W, Ma S, Sun X, Gu R J, Faraone L and Lei W 2023 Structural properties and defect formation mechanisms in MBE-grown HgCdTe on InSb (211)B substrates *J. Appl. Phys.* **134** 115303
[78] Jaime-Vasquez M, Martinka M, Stoltz A J, Jacobs R N, Benson J D, Almeida L A and Markunas J K 2008 Plasma-cleaned InSb (112) B for large-area epitaxy of HgCdTe sensors *J. Electron. Mater.* **37** 1247–54

Chapter 4

Heteroepitaxial growth of HgCdTe on lattice-mismatched two-dimensional substrates

In chapter 3, we discussed the hetero-epitaxial growth of CdTe and HgCdTe materials on various lattice-mismatched substrates including Si, Ge, GaAs, and GaSb. The main constraint for growing high-quality CdTe and HgCdTe materials on these alternative substrates is the large lattice constant mismatch between CdTe/HgCdTe and substrates which leads to a high density of threading dislocations in the HgCdTe materials and thus deteriorates the performance of resultant IR detectors, especially LWIR detectors. Therefore, it is critical to reduce the threading dislocation density in CdTe/HgCdTe materials. In parallel with those dislocation reduction techniques as discussed in chapter 3, this chapter will introduce an alternative epitaxial growth technique—Van der Waals epitaxy to relax the stringent constraint of lattice constant matching of heteroepitaxial growth, and thus achieve high-quality CdTe and HgCdTe materials on alternative substrates.

4.1 Introduction to Van der Waals epitaxy of CdTe and HgCdTe materials on two-dimensional substrates

As discussed in chapter 3, the large lattice mismatch between CdTe/HgCdTe layer and alternative substrate constitutes the biggest challenge to the MBE growth of CdTe and HgCdTe on alternative substrates such as Si, Ge, GaAs, and GaSb. The large lattice mismatch generates a large number of misfit dislocations which form threading dislocations to propagate into the top HgCdTe layers. This results in the high EPD numbers (mid-10^6 cm^{-2} to low-10^7 cm^{-2}) observed in the CdTe and HgCdTe layers grown on these alternative substrates. Although various techniques have been proposed to annihilate and reduce the threading dislocations in the CdTe and HgCdTe layers, the EPD numbers are still much higher

 © IOP Publishing Ltd 2025. All rights, including for text and data mining (TDM), artificial intelligence (AI) training, and similar technologies, are reserved.

than that (low-10^4 cm^{-2} to low-10^5 cm^{-2}) in HgCdTe grown on lattice-matched CdZnTe substrates, and higher than that (5×10^5 cm^{-2}) required for making high-performance LWIR HgCdTe detectors. Therefore, new alternative epitaxial growth methods are required to grow high-quality CdTe and HgCdTe layers on lattice-mismatched alternative substrates. As discussed in chapter 2, two-dimensional (2D) Van der Waals epitaxy (vdWE) provides a promising alternative growth method for growing heterostructures without the limitation of lattice mismatch.

2D vdWE is a growth technique where the epitaxial layer is deposited onto a 2D layered substrate such as graphene and mica, which are materials composed of atomically thin layers held together by weak interlayer Van der Waals (vdW) forces. These materials exhibit no dangling bonds on their surfaces, providing atomically smooth, chemically inert surfaces ideal for vdW epitaxy. Unlike traditional semiconductor epitaxy, which relies on strong dangling bonds and requires close lattice matching between the substrate and the film to minimize strain and defects, the stringent requirement of lattice constant matching for high-quality heterostructure growth can be relaxed for vdWE growth since strong chemical bonding is not present at the interfaces, as shown in figure 2.4. The weakened vdW interaction at the hetero-interface significantly reduces the strain that would typically arise from lattice mismatch in traditional epitaxial growth, thus reducing the generation of misfit dislocations at the heterostructure interfaces [1, 2]. This offers a unique opportunity to grow high-quality CdTe and HgCdTe layers on 2D layered substrates without the limitation of lattice constant matching. Another essential feature of vdWE growth is that the epitaxial layers can be easily lifted off the substrates to form free-standing thin films due to the weak vdW force between the layers [3]. This enables the easy, damage-free removal of substrates from the epitaxial layers and transfer among various substrate carriers, and thus allows the fabrication of various flexible detector devices. Furthermore, some post-released substrates (such as graphene-coated substrates) can be reused for multiple growths and transfer cycles [4], which will lower the cost of material and device fabrication. Thus, the vdWE approach will not only enable the fabrication of high-quality CdTe and HgCdTe layers on lattice mismatched 2D substrates, but broaden their applications in various advanced electronic and optoelectronic devices.

As discussed in chapter 2, generally the vdWE of HgCdTe on 2D substrates can be categorized into two primary approaches:

Direct vdWE: For this technique, CdTe and HgCdTe layers are deposited directly onto 2D substrates (such as mica), taking advantage of the weak interfacial vdW bonding to relax interfacial strain and minimize defect formation to achieve high-quality epitaxy [2].

Remote vdWE: For this technique, HgCdTe is deposited onto 2D material-coated rigid three-dimensional (3D) substrates (such as graphene-coated CdZnTe substrates), allowing the system to inherit the surface orientation of the underlying 3D substrate and enabling spontaneous strain relaxation by modifying the surface energy through the 2D coating to facilitate subsequent epitaxial growth [5].

It should be noted that mica and graphene are known to remain pristine during epitaxy without dissolving into the substrate or epilayer due to their high thermal stability, making them ideal substrates of choice for the vdWE growth of 3D materials. To date, noteworthy studies have been undertaken on the epitaxial growth of 3D materials on 2D substrates include the vdWE of GaN, GaAs, CdTe, CdSe, CdS, ZnTe, etc [4–14]. Both direct and remote vdWE approaches offer promising strategies to overcome the limitations of conventional heteroepitaxy and open new possibilities for growing high-quality CdTe and HgCdTe materials for making high-performance IR detectors, especially LWIR ones.

4.2 Heteroepitaxial growth of CdTe and HgCdTe on graphene substrates

4.2.1 Introduction of two-dimensional graphene substrates

Graphene, a 2D material consisting of a single layer of carbon atoms arranged in a honeycomb lattice, has become one of the most intensively studied materials in the field of materials science and nanotechnology. Since its isolation by Novoselov and Geim in 2004 using mechanical exfoliation (Scotch tape method) [15], graphene has attracted global attention due to its extraordinary mechanical strength, ultrahigh carrier mobility, high optical transparency, excellent thermal conductivity and stability, and large specific surface area. This pioneering discovery led to the Nobel Prize in Physics in 2010.

Apart from its intrinsic properties, graphene offers unique opportunities as a substrate or interfacial layer for epitaxial growth. One of its most important advantages is its atomically flat and chemically inert surface, which is free of dangling bonds. When undertaking vdWE, the weak vdW interaction between graphene and an overgrown epilayer relaxes lattice-matching constraints, allowing high-quality crystalline films to be deposited even on highly mismatched substrates. The graphene layer can thus serve as an effective compliant platform for strain relaxation and defect suppression. As a result, graphene has been proposed and demonstrated as an ideal platform for the vdWE of various semiconductors, including II–VI and III–V compounds, and transition metal dichalcogenides (TMDs) [16, 17]. In this section, we will mainly focus on the vdWE growth of 3D semiconductor materials on 2D graphene substrates.

Nowadays, graphene is readily available and scalable. Commercial suppliers offer both monolayer and multilayer graphene films on various substrates (SiO_2/Si, quartz, copper foils, etc) at a reasonable cost, depending on the desired quality and substrate size [18]. Large-area monolayer graphene can also be grown on copper foils via chemical vapor deposition (CVD), which can then be transferred onto target substrates such as SiO_2/Si and others. Apart from CVD growth, the graphitization process on silicon carbide (SiC) substrates provides another popular way to achieve uniform, large-area monolayer or few-layer graphene films without the need for transfer printing. [4, 14, 19, 20]. Note that the crystalline quality of graphene is crucial for the subsequent vdWE growth. Single-crystalline graphene is often required to guide the oriented growth of overlying films and to reduce the introduction of grain boundaries and nucleation-related defects. The crystallinity, domain size, and surface cleanliness of graphene significantly influence the nucleation dynamics, grain alignment, and defect density of the epitaxial layers grown on it. Therefore, people have to ensure that graphene thin film is of single crystalline quality, and the surface of graphene thin film is clean without any contaminants. Various techniques can be used to achieve this, including using single-crystal growth templates [12, 13, 21, 22], and post-transfer annealing.

4.2.2 General Van der Waals epitaxial growth on two-dimensional graphene substrates

One of the earliest and most influential demonstrations of single-crystalline epitaxy via vdWE on graphene was reported by Kim *et al* in 2014 [4], where they successfully grew single-crystalline GaN layers on epitaxial graphene formed on 4H–SiC (0001) substrates. This work provided a fundamental proof-of-concept that high-quality 3D semiconductors could be grown on 2D graphene without traditional lattice-matching constraints, marking a major milestone in the field of vdWE. This vdWE process began with the preparation of single-crystalline epitaxial graphene, which includes multistep annealing and SiC graphitization in at Ar atmosphere at 1575 °C. The resulting graphene was highly ordered and epitaxially aligned with the underlying SiC crystal, offering an atomically flat, single-crystalline surface ideal for vdWE. Following graphene formation, GaN films were deposited directly onto the graphene layer using MOCVD. Remarkably, the GaN nucleated in a highly ordered manner, resulting in single-crystalline GaN epitaxial layers with low defect density, smooth surface morphology, and crystalline quality comparable to conventional GaN grown directly on SiC.

Figure 4.1(a) schematically illustrates the overall process of GaN growth on epitaxial graphene, including graphene synthesis, vdWE growth, and film transfer [4]. Figures 4.1(b)–(e) present a range of structural characterizations, including high-resolution XRD (HRXRD), scanning electron microscopy (SEM), AFM, and high-resolution transmission electron microscopy (HRTEM) [4], all of which confirm the high crystalline quality and smooth surface of the GaN film. An important outcome of this study was the demonstration that the resulting GaN films could be lifted off from the graphene/SiC substrate and transferred to arbitrary substrates. This was achieved without damaging the crystal quality of the GaN, illustrating another benefit of vdWE: the potential for scalable, transferable epitaxial layers. Such capabilities are especially useful for developing flexible or heterogeneous device architectures where traditional substrate constraints are limiting.

One of the most significant challenges in realizing high-quality epitaxial growth of 3D materials on graphene lies in the inherently suppressed nucleation of adatoms at the heterostructure interface. Pristine graphene, with its atomically smooth and chemically inert surface, lacks reactive sites or dangling bonds, which are typically

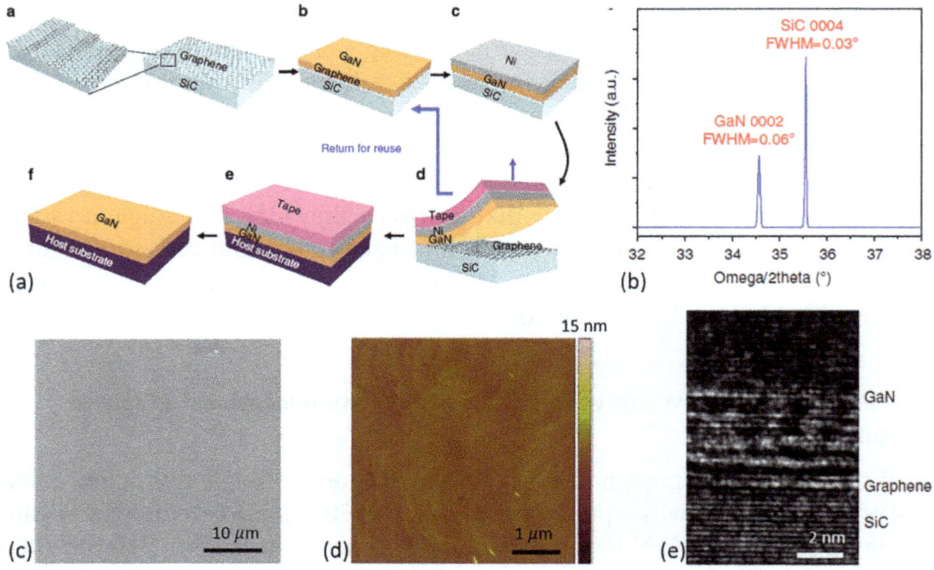

Figure 4.1. (a) Schematic diagram of the direct vdWE process followed by lift-off and transfer of the GaN epitaxial film; (b) HRXRD $\omega - 2\theta$ scan confirms the crystalline quality of the GaN layer; (c) plan-view SEM image reveals a uniform and continuous surface morphology; (d) AFM image indicates an ultra-smooth surface with a RMS roughness of 3 Å; (e) cross-sectional HRTEM image shows well-aligned crystal lattices, demonstrating coherent epitaxial growth on the graphene substrate. Reproduced from [4] with permission from Springer Nature.

required for adatom binding and nucleation [4, 7, 9–11, 23–25]. As a result, most 3D materials exhibit poor wettability on graphene surfaces, inhibiting uniform nucleation and often resulting in island-like or polycrystalline growth. Since the initial nucleation phase strongly influences film morphology, crystalline orientation, defect density, and interfacial adhesion [26], overcoming this nucleation barrier is crucial to realizing the full potential of vdWE on graphene. To address this issue, interfacial engineering strategies have been developed to promote nucleation and improve epitaxial quality. These strategies include: (1) insertion of interfacial buffer layers, (2) creation of nucleation or defective sites on the graphene surface, and (3) modulation of growth temperature. Since growth temperature is a growth parameter to optimize anyway, the following discussion will focus on strategies (1) and (2). They are:

Interfacial buffer layers: For the vdWE of GaN on graphene several buffer layer methods have been proposed to enhance GaN nucleation and growth, such as using ZnO nanowalls [7, 27–29], AlN buffer layer [14, 19, 20, 30–32], and AlGaN buffer layers [23]. While these buffer layers can improve nucleation coverage and surface morphology, the crystalline quality of GaN films grown by such approaches often remains inferior to that achieved on conventional substrates such as sapphire or SiC.

Creation of nucleation or defective sites: In addition to buffer layers, surface engineering of graphene itself has emerged as an effective strategy to promote nucleation. Two-step growth techniques, in which a low-temperature nucleation layer is first deposited and then followed by a standard-temperature epitaxial growth step, have shown promising results. Kim *et al* demonstrated that two-step MOCVD growth enabled the formation of GaN films on graphene with structural and optical quality comparable to GaN grown on standard SiC substrates, indicating that this method may be one of the most viable routes for achieving high-quality epitaxy on graphene [4]. Apart from this low-temperature nucleation layer, UV treatment has also been used to intentionally introduce defective or functionalized sites on graphene, thereby enhancing nucleation [10].

A significant advancement in the field of vdWE emerged with the concept of remote epitaxy, first introduced by Kim *et al* [5]. Unlike conventional vdWE, where epitaxial films are grown directly on 2D materials such as graphene or mica, remote epitaxy involves growing epitaxial layers on a 2D material-coated 3D substrate, typically using a monolayer of graphene as the interfacial layer. This approach leverages the fact that monolayer graphene is atomically thin and electronically transparent, allowing the electrostatic potential of the underlying substrate to penetrate through the graphene and guide the crystallographic orientation of the overgrown film. Due to the weak vdW interaction at the graphene-film interface, the overlayer can be detached from the substrate after growth, enabling non-destructive release and transfer of single-crystalline thin films. This not only allows for substrate reuse but also facilitates the fabrication of flexible, transferable, and potentially curved devices. Kim *et al* demonstrated this concept by growing high-quality, single-crystalline GaAs on monolayer graphene-coated GaAs (001) substrates using a two-step MOCVD process [5]. As illustrated in figure 4.2, the resulting GaAs film exhibited excellent structural quality [5], confirmed by electron back scatter diffraction (EBSD) mapping and HRXRD φ-scan, with no in-plane rotational domains. The study showed that the presence of a single layer of graphene was insufficient to screen the potential field of the underlying substrate, allowing the overgrown film to align with the underlying substrate.

Subsequent work by Kong *et al* systematically investigated the influence of graphene thickness on the effectiveness of remote epitaxy [33]. Their findings revealed that the substrate's electrostatic potential becomes fully screened when the graphene thickness exceeds three monolayers, preventing the overgrown film from adopting the crystallographic orientation of the substrate and thus reverting the process to conventional vdWE. This indicates that remote epitaxy is only feasible when the graphene thickness is limited to fewer than three atomic layers, ideally a monolayer, to maintain crystallographic alignment with the underlying substrate. Further developments in remote epitaxy have extended beyond GaAs to other material systems. Notably, Journot *et al* and Badokas *et al* demonstrated remote epitaxial growth of GaN on monolayer graphene-coated substrates, emphasizing the critical role of the nucleation mechanism in achieving high crystalline quality [34, 35]. Their findings suggested that while remote epitaxy enables lattice guidance from the substrate, the nucleation process must still be carefully engineered through

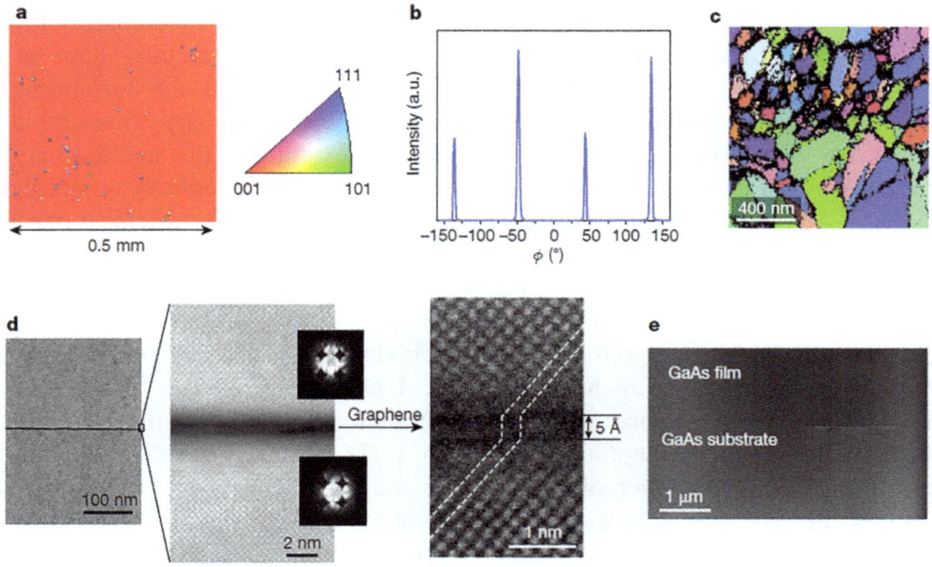

Figure 4.2. Characterization of GaAs films grown on monolayer graphene-coated GaAs (001) substrates via remote epitaxy. (a) EBSD map of the released GaAs layer reveals single-crystalline orientation; (b) HRXRD φ-scan confirms the zinc blende crystal structure with no in-plane rotation; (c) EBSD map of GaAs grown on graphene without H_2 annealing post-transfer shows degraded crystallinity; (d) high-resolution-scanning transmission electron microscopy (STEM) image demonstrates excellent lattice alignment across the monolayer graphene interface; and (e) low-angle annular dark-field (LAADF) STEM image indicates the absence of dislocations, confirming high crystalline quality. Reproduced from [5] with permission from Springer Nature.

temperature control, surface treatment, and buffer layers to ensure uniform film growth and defect minimization.

All the above direct and remote vdWE growths provide valuable guidance on realizing high-quality vdWE growth of CdTe and HgCdTe on 2D graphene substrates.

4.2.3 Van der Waals epitaxial growth of HgCdTe on two-dimensional graphene substrates

Van der Waal epitaxy of CdTe on graphene has emerged as a promising approach to realize high-quality II–VI semiconductors on lattice-mismatched substrates. As CdTe serves as a key template for HgCdTe growth, its successful vdWE integration on graphene can lay a solid foundation for future growth of HgCdTe on graphene. One of the earliest investigations was reported by Jung *et al* in 2013 [10], where CdTe thin films were grown on graphene-buffered SiO_2/Si substrates using the close-spaced sublimation (CSS) technique. While the resulting films remained polycrystalline with noticeable surface roughness and grain boundaries, the study demonstrated that graphene could serve as a functional buffer layer to promote CdTe nucleation and film continuity. Moreover, they found that defect engineering of the graphene layer, specifically through UV treatment, could enhance nucleation uniformity and enable more continuous and uniform film growth.

Building on this concept, Mohanty and co-workers made significant advancements in the following years. Initially, in 2016 they reported the vdWE of CdTe on monolayer graphene transferred onto SiO_2/Si substrates, despite the substantial lattice mismatch of 46% between CdTe and graphene [11]. In 2019, they improved film quality by using single-crystalline graphene transferred onto amorphous SiO_2/Si substrates, and the resultant CdTe films exhibited a single-crystalline structure composed of twin domains [36]. Their first-principle calculations and experimental observations confirmed that the weak vdW interface facilitates strain relaxation and prevents the formation of misfit dislocations. These studies highlight the critical role of both graphene crystallinity and surface preparation in enabling high-quality epitaxial CdTe growth via vdWE. The transition from polycrystalline to quasi-single-crystalline CdTe films demonstrates the critical role of the graphene template in dictating the film's structural properties.

Despite the encouraging progress for CdTe, there are currently no open reports on HgCdTe grown on graphene. However, given the same zinc blende crystal structures of CdTe and HgCdTe, the foundational work on CdTe provides a valuable guidance. At UWA, we have proposed a remote vdWE strategy for growing HgCdTe by using monolayer single-crystalline graphene-coated CdZnTe substrates which will be implemented in the near future. This method leverages the concept that monolayer graphene is sufficiently thin to allow the electrostatic potential of the underlying CdZnTe to influence the crystallographic orientation of the overgrown HgCdTe film, enabling epitaxy despite the presence of the intervening graphene layer. This technique has been previously validated for III–V semiconductors such as GaAs and GaN [5, 35], where remote epitaxy through graphene allowed the overgrown films to inherit the substrate orientation while remaining detachable due to the weak vdW interface.

To make graphene-coated CdZnTe substrates, a monolayer of graphene will be purchased from established commercial suppliers of graphene and then transferred onto the fresh surface of CdZnTe substrates. Initially, graphene transfer will be outsourced to these commercial vendors to accelerate development, while in parallel, an in-house transfer process will be developed. Building on prior experience in transferring 2D materials such as Bi_2Te_3, Bi_2Se_3, and SnTe [37, 38] in our previous work, a Ni stressor-based exfoliation method will be employed. Firstly, a Ni stressor layer will be deposited on the surface of commercial graphene on a carrier substrate and then the Ni/graphene will be exfoliated from the carrier substrate using a thermal-releasing tape handler. The Ni/graphene will be immediately transferred onto the CdZnTe substrate which should be cleaned with a wet etch (Br in methanol) to remove oxides on the substrate surface. Then, TFB nickel etchant will be used to remove the Ni and leave the graphene/CdZnTe substrate for loading into the MBE system. Similar methods have been reported for remote vdWE epitaxial growth of single-crystalline III–V films on graphene-coated III–V substrates [5, 39]. Raman spectroscopy will be used to assess graphene integrity after transfer. This approach avoids wet transfer damage, minimizes residue, and maintains graphene continuity, which is critical for subsequent vdWE growth. Once optimized, this system offers the unique potential to combine remote vdWE alignment with clean layer transfer. Note that the use of CdZnTe as the underlying substrate ensures compatibility with existing HgCdTe MBE growth protocols, facilitating a smoother transition from a conventional to a remote vdWE process.

Despite this plan, there are a number of challenges to be addressed before achieving high-quality 3D HgCdTe materials on 2D graphene substrates. These include:

(1) *Crystalline quality and surface quality of graphene:* For both direct and remote vdWE, it is critical to have graphene with high crystallinity quality and surface quality. Single-crystalline graphene enables uniform interfacial interaction and consistent crystallographic orientation, which are essential for achieving high-quality epitaxial growth with minimal defects. The surface should be smooth and clean without any wrinkling, tearing, or contamination.

(2) *Nucleation density and surface inertness:* Pristine graphene is chemically inert and lacks dangling bonds, which suppresses the nucleation of adatoms during epitaxial growth. Since nucleation is a critical stage in determining the final film's crystallinity, morphology, and defect density, low nucleation density represents a major challenge. Those approaches proposed to increase the nucleation density for the vdWE of GaN on graphene can be applied for the vdWE of HgCdTe on graphene.

(3) *Crystal orientation and twin domain:* Even with high-quality graphene, achieving single-crystalline epitaxial layers remains challenging due to the absence of strong in-plane chemical bonding at the graphene interface. As discussed previously, materials such as CdTe frequently exhibit multiple in-plane orientations or twin domains, which should be avoided.

4.3 Heteroepitaxial growth of CdTe and HgCdTe on mica substrates

4.3.1 Introduction of two-dimensional mica substrates

Mica has recently attracted attention as a potential 2D substrate for the hetero-epitaxial growth of semiconductors, including CdTe and HgCdTe. Among various types of micas, fluorphlogopite mica [$KMg_3(AlSi_3O_{10})F_2$] is a synthetic layered mica that is commercially available and widely used for research purposes due to its atomically flat cleavage plane, excellent thermal stability, optical transparency, and low cost. Mica has a flexible layered structure where the first few layers can be easily cleaved and peeled off, exhibiting no dangling bonds at its smooth surface with strong in-plane chemical bonding. Mica substrates usually exhibit (001) single crystalline orientation. This provides a great possibility to grow high-quality single-crystalline epitaxial layers on mica for further device fabrication [9].

Since mica can be easily exfoliated to produce atomically smooth surfaces, it provides a convenient and cost-effective platform for exploratory epitaxial growth studies. However, mica presents a high-energy surface that readily adsorbs water, organic contaminants, and gases from the atmosphere, and thus this surface needs proper treatment before being loaded into the MBE growth chamber for growing CdTe and HgCdTe layers. Our UWA preliminary results indicate that a thermal cleaning at ∼400 °C can lead to a 2D surface with atomic flatness as observed by the RHEED pattern, providing a smooth growth front for the subsequent epitaxial growth of II–VI thin films. Ideally, the exfoliation and transfer of mica substrates should be undertaken in an inert environment (e.g., a N_2- or Ar gas-filled glove box) integrated with the MBE sample loading system, which ensures an O_2- and moisture-free environment, and thus preserves surface cleanliness for high-quality heteroepitaxy.

Note that mica has the potential to become an industry substrate for making IR detectors. Mica substrates are available in large wafer sizes (up to 4 inches), offer high IR transparency (>65% in SWIR, >80% in MWIR, and >50% in LWIR ranges), and are significantly more cost-effective (∼$54 per 2-inch wafer) than lattice-matched CdZnTe substrates. These attributes make mica an attractive candidate substrate for growing IR materials and detectors.

4.3.2 General Van der Waals epitaxial growth on two-dimensional mica substrates

In recent years, 2D mica substrates, particularly synthetic fluorphlogopite mica, have emerged as promising platforms for the heteroepitaxial growth of a wide range of semiconductor materials via vdWE including MoO_2 [40], Al-doped ZoN [41], Ga_2O_3 [42], GaN [43], CdTe [44, 45], and CdSe [3], ZnTe [46], etc.

A notable relevant breakthrough was reported by Lian *et al* in 2019, which successfully demonstrated the direct MBE growth of single-crystalline CdTe layers on mica [47]. Figure 4.3 shows the material characterization results of the CdTe grown on mica. The 120 nm-thick CdTe/mica heterostructure exhibited a remarkably narrow XRD FWHM of 0.05°, surpassing the crystalline quality of similar

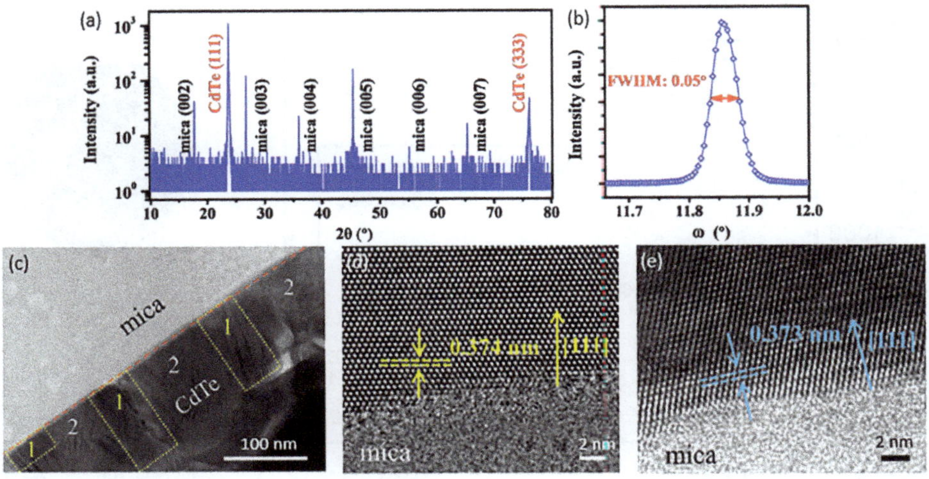

Figure 4.3. Material characterization of CdTe grown on mica. (a) High-resolution XRD $\omega - 2\theta$ scan and (b) rocking curve confirm the single-crystalline nature of the CdTe film with a narrow FWHM of 0.05°; (c) cross-sectional TEM image reveals distinct structural regions; (d, e) high-resolution TEM images corresponding to regions 1 and 2 in (c), showing well-defined lattice fringes indicative of high crystalline quality. Reproduced from [47] John Wiley & Sons. © 2019 WILEY-VCH Verlag GmbH & Co. KGaA, Weinheim.

CdTe layers grown on graphene substrates [10, 11]. HRTEM analysis confirmed well-aligned lattice fringes at the interface, and fabricated photodetectors based on this heterostructure showed high responsivity, detectivity, and rapid response times. Furthermore, the device retained excellent mechanical robustness under bending, highlighting the potential of CdTe/Mica systems for flexible optoelectronic applications.

Our research group at UWA has also initiated a preliminary investigation into the direct vdWE-MBE growth of CdTe and CdSe buffer layers on 2D mica substrates, aiming to lay the groundwork for growing high-quality HgCdTe and HgCdSe IR materials. As shown in figure 4.4(a), XRD rocking curve measurements of a CdTe epilayer grown on mica exhibited a relatively narrow FWHM (~115 arcsec), indicative of good crystalline quality despite the significant lattice mismatch. Moreover, the feasibility of a damage-free lift-off process was demonstrated, as shown in figure 4.4(b). The wafer-scale (2-inch) CdTe layer was effectively detached from the mica substrate after immersion in deionized water with mild ultrasonic agitation, validating the weak vdW interaction at the interface. In parallel, our group also explored the vdWE of CdSe on mica [3], which demonstrated large-area, flexible, and highly crystalline 130 nm-thick CdSe films grown via vdWE. This study confirmed that mica substrates enable oriented CdSe growth with strong in-plane alignment and minimal grain boundaries, resulting in thin films with excellent optical and mechanical performance. A flexible photoconductive device was fabricated with good device performance, enabling high-resolution full-color imaging across the visible spectral range.

Similar to graphene substrate, the chemically inert and dangling-bond-free surface of mica substrate also presents a significant challenge for the nucleation of

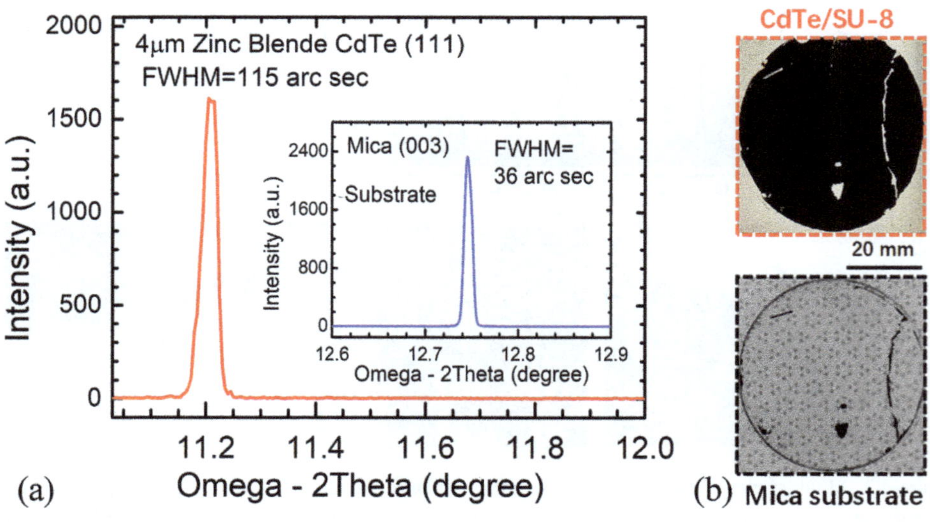

Figure 4.4. (a) XRD rocking curve for a CdTe epilayer grown on Mica; (b) photo of CdTe epilayer and mica substrate after lift-off in DI water with a low level of ultrasonic vibration. The inset of (a) shows the XRD rocking curve of a mica substrate.

3D semiconductors, often resulting in low nucleation density and poor initial film quality. To overcome this, several interfacial engineering strategies have been proposed and studied, including:

(1) *Buffer layer:* Wang *et al* demonstrated the use of a SnSe buffer layer for the MBE growth of SnS on mica [48]. The introduction of the SnSe buffer improved lattice alignment, enhanced film texture, and significantly reduced trap state density in the SnS epilayer, indicating the effectiveness of a chemically and structurally compatible buffer in promoting nucleation and crystalline quality.

(2) *Seeding layer:* Lu *et al* reported the chemical solution deposition of VO_2 films on mica, where a thin VO_2 seeding layer was first deposited to promote uniform nucleation [49]. This technique significantly improved the grain size, reduced defect density, and enhanced overall crystallinity.

All these interfacial engineering strategies can be extended for growing high-quality CdTe and HgCdTe on mica substrates with further effort on them.

4.3.3 Van der Waals epitaxial growth of HgCdTe on two-dimensional mica substrates

While significant progress has been made in the vdWE of CdTe and other materials on mica, there is only limited information on the vdWE of HgCdTe on mica. To date, only our research group at UWA has reported two studies on the successful vdWE growth of HgCdTe on mica substrates.

In 2023, Pan *et al* demonstrated the direct MBE growth of 5.5 μm-thick MWIR HgCdTe (111) thin films on transparent mica substrates via vdWE [50]. The resulting heterostructures exhibited a XRD FWHM of approximately 306 arcsec as shown in figure 4.5(b), indicative of moderately high crystalline quality. High-resolution cross-sectional TEM imaging in figure 4.5(c) revealed the presence of twin domains originating at the film/substrate interface, which contributed to the structural imperfections and limited device performance. Nevertheless, IR transmission spectra measured at 300 K showed excellent optical transparency, and simulations based on the experimental data confirmed the accuracy of the extracted layer structure, as shown in figure 4.5(a). Figure 4.5(d) indicates that the photoconductive detectors fabricated from these films exhibited promising temperature-dependent performance, with responsivities of $\sim 110\, V \cdot W^{-1}$ at 80 K and $\sim 8\ V \cdot W^{-1}$ at 300 K. Notably, the weak interfacial bonding between HgCdTe and mica enabled a clean, damage-free lift-off process, supporting the fabrication of large-area, transferable, free-standing thin-film devices, which is an essential step towards flexible IR optoelectronics.

Building on these initial findings, a follow-up study by Ma *et al* in 2024 further investigated the vdWE growth of HgCdTe on Mica, focusing on optimizing growth parameters to improve the film quality [51]. The study highlighted the critical importance of substrate preparation and precise control of growth conditions, particularly substrate temperature and Hg flux ratio, in minimizing defect formation and enhancing film uniformity. In this work, 3.5 μm-thick MWIR HgCdTe thin film

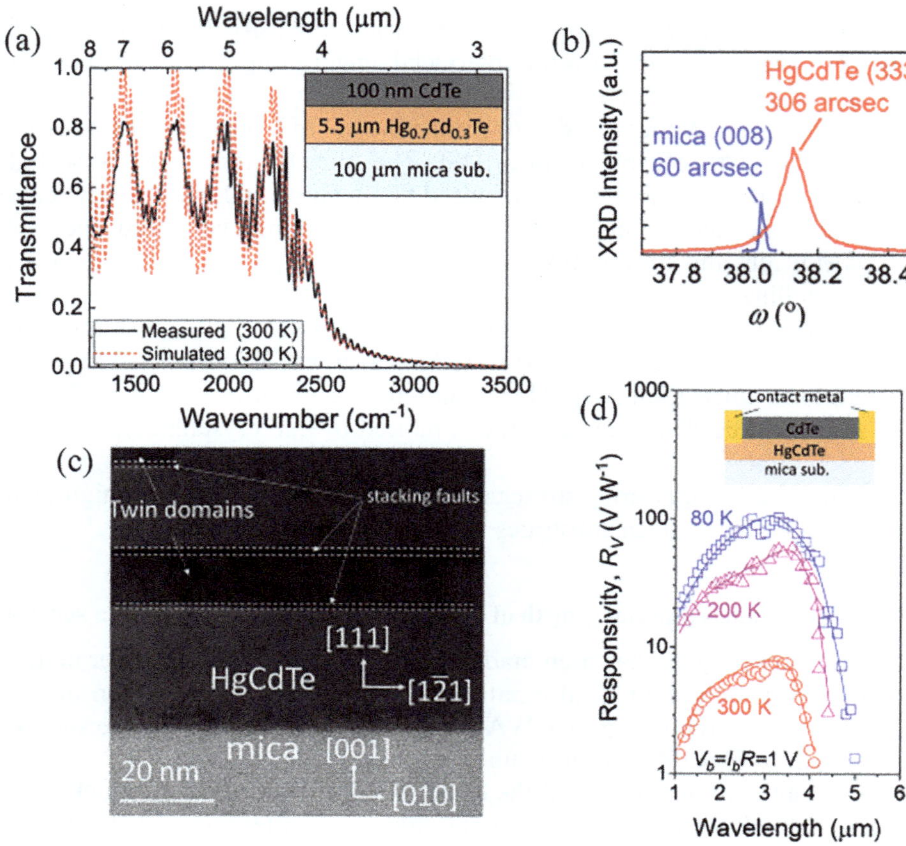

Figure 4.5. (a) Experimental and simulated infrared transmission spectra of a representative vdWE-MBE-grown HgCdTe/mica sample measured at 300 K. The inset shows the extracted layered structure used for modeling. (b) High-resolution XRD ω-scan of the HgCdTe (111) epilayer, indicating a FWHM of ~306 arcsec. (c) Cross-sectional HRTEM image showing twin regions near the HgCdTe/mica interface, illustrating defect formation. (d) Temperature-dependent photoresponse spectra of fabricated HgCdTe detectors operated at 80 K, 200 K, and 300 K, revealing broadband responsivity across the MWIR range. Reproduced from [50] John Wiley & Sons. © 2022 The Authors. Advanced Materials Interfaces published by Wiley-VCH GmbH.

layers were grown under both optimized and non-optimized conditions. Figure 4.6 compares the structural and morphological characteristics of the resulting films. SEM, AFM, and cross-sectional TEM images (figures 4.6(a)–(c)) of the sample grown under optimized parameters showed smooth surface morphology and low dislocation densities, despite the presence of twin domains. In contrast, the sample grown under non-optimal conditions (figures 4.6(d)–(f)) displayed rougher surfaces and significantly higher defect densities, leading to degraded crystal quality. These results highlight that careful tuning of growth parameters can substantially improve material quality. However, the persistent issue of twinning remains a key challenge requiring further investigation.

As discussed above, the material quality of HgCdTe grown on mica substrates still cannot compete with that grown on traditional CdZnTe substrates. There are a

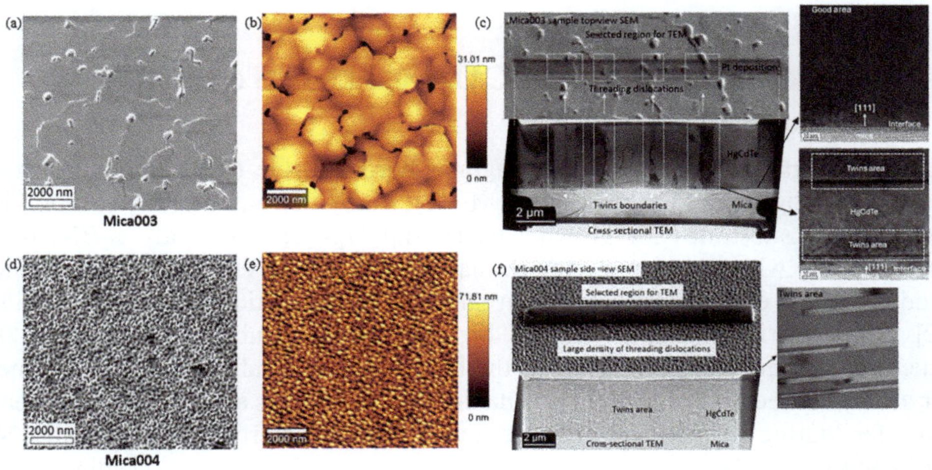

Figure 4.6. (a) SEM image, (b) AFM surface morphology, and (c) cross-sectional TEM of HgCdTe films grown on mica under optimized growth temperature and Hg flux ratio, demonstrating smooth surface and reduced defect density; (d) SEM image, (e) AFM, and (f) cross-sectional TEM of HgCdTe films grown under non-optimal growth conditions, showing increased surface roughness and higher defect density. These comparisons highlight the critical influence of growth parameters on the structural quality of vdWE-grown HgCdTe epilayers. Reproduced from [51]. CC BY 4.0.

number of challenges to be addressed before achieving high-quality HgCdTe materials. These challenges include:

(a) *Surface contamination during mica substrate preparation:* As with graphene substrates, it is critical to ensure a clean and atomically flat surface of mica substrates prior to epitaxy. Mica presents a high surface energy that makes it highly susceptible to adsorption of moisture, organic contaminants, and airborne particulates during handling in ambient conditions. These surface defects can significantly degrade nucleation behavior and reduce the crystalline quality of the subsequently grown films. Thus, approaches must be undertaken to minimise the potential contaminants such as thermal annealing in a vacuum, inert-gas-filled glove box for substrate exfoliation and preparation.

(b) *Low nucleation density:* The chemically inert, atomically smooth, and dangling-bond-free surface of mica poses a significant challenge for adatom nucleation, often resulting in low nucleation density and discontinuous film growth during the initial stages of vdWE. To address this, various interfacial engineering strategies should be studied such as buffer layers, seeding layers, and others.

(c) *Twin formation and crystal defects:* As discussed before, twin defects were frequently observed for CdTe and HgCdTe grown on mica. To improve the crystal quality, various growth parameters as well as other methodologies should be studied and optimized to effectively suppress and eliminate twin defects. Apart from these, other dislocation reduction techniques as discussed in chapter 3 can also be studied, such as thick buffer layers, strained superlattice DFLs, and others.

4.4 Heteroepitaxial growth of CdTe and HgCdTe on other two-dimensional substrates

Apart from graphene and mica, hexagonal boron nitride (h-BN) and transition metal dichalcogenides (TMDs) such as $MoTe_2$, MoS_2, WSe_2, and WS_2 also provide potential 2D substrates for growing CdTe and HgCdTe materials. h-BN, often referred to as 'white graphene' is an insulating 2D material characterized by a wide bandgap (~5.9 eV), an atomically flat surface, and exceptional chemical stability [52]. These attributes make h-BN a highly promising substrate for vdWE. Its structural similarity to graphene and the low interlayer binding energy (~2.0 meV per atom) between h-BN sheets also enable it to function as an intermediate release layer, facilitating clean lift-off processes in a variety of heteroepitaxial systems [24]. h-BN flakes can be mechanically exfoliated from bulk crystals, yielding clean, atomically thin layers suitable for laboratory-scale studies [53]. For scalable applications, large-area h-BN films can be synthesized via CVD on catalytic substrates such as copper or nickel foils. h-BN can also be directly grown using MOCVD with precursors such as triethylborane (TEB) and ammonia (NH_3) [54–59]. In contrast to h-BN, TMD monolayer possesses a unique monolayer structure composed of hexagonal lattices [24]. Their chemical robustness and relatively high thermal stability make them compatible with high-temperature epitaxial growth processes. High-quality TMD monolayers or few-layer flakes are typically obtained through mechanical exfoliation from bulk crystals, which are often synthesized via CVD or MOCVD [60–64].

To date, the majority of vdWE studies on h-BN and TMD monolayer substrates focus on III-nitride materials, particularly GaN, AlN, and InGaN, as well as their corresponding optoelectronic devices [54–59]. For example, wafer-scale InGaN/GaN multiquantum well (MQW) structures have been successfully grown on h-BN and subsequently transferred onto arbitrary substrates, paving the way for the realization of flexible light-emitting diodes (LEDs) and other transferable device architectures [54, 59]. Another example, strain-free, single-crystalline GaN islands have been grown on mechanically exfoliated MoS_2 and WS_2 flakes via MOCVD [63]. Despite the progress on III-nitride materials, there is not much information available on the vdWE of CdTe and HgCdTe on these h-BN and TMD monolayer substrates. One representative study demonstrated the MBE growth of CdTe quantum wells on exfoliated h-BN flakes [53]. The growth was conducted at varying substrate temperatures, and an optimal growth temperature of approximately 220 °C was identified which is around 100 °C lower than the typical growth temperature required for CdTe on conventional 3D substrates. This substantial decrease in optimal growth temperature reflects the fundamentally different growth dynamics on 2D layered substrates. The successful formation of quantum wells at reduced temperatures also suggests that h-BN may offer a thermally compatible platform for integrating II–VI materials onto flexible or temperature-sensitive substrates. Apart from CdTe on h-BN substrates, several II-VI materials including

CdTe, CdS, and InSe, have been explored for the vdWE on TMD monolayer substrates such as $MoTe_2$ and WSe_2 [60–62]. While these studies highlight the potential of TMD monolayers as lattice-mismatched, chemically stable platforms for 3D material integration, the resulting films have generally exhibited suboptimal crystallinity. Common issues include polycrystalline films, the formation of 3D island-like clusters, and the presence of twin domains, all of which degrade the structural and electronic properties of the grown materials [60–64].

As discussed in the previous section, CdTe has been experimentally grown on layered 2D substrates such as h-BN and TMD monolayer including $MoTe_2$ and WSe_2. These early studies provide valuable insight into the feasibility of vdWE for II–VI semiconductors on 2D materials beyond graphene and mica. However, the outcomes remain far from ideal. The resulting CdTe films often suffer from polycrystalline microstructures, 3D island-like growth modes, and orientation disorder, which significantly compromise their suitability for high-performance optoelectronic applications. So far, there is no report on the vdWE of HgCdTe on h-BN and TMD monolayer substrates. Therefore, significant effort is needed to address the current problems and challenges before achieving high-quality vdWE of CdTe and HgCdTe on h-BN and TMD monolayer substrates. These problems and challenges include:

(a) *High-quality substrates:* One of the primary challenges to achieving high-quality vdWE is the lack of large-area, defect-free, single-crystalline h-BN and TMD monolayer substrates. Most existing studies rely on exfoliated flakes or synthesized films, which often contain grain boundaries, wrinkles, or non-uniform thicknesses [54–64]. These structural imperfections can disrupt the epitaxial alignment and compromise the crystallinity of the overgrown CdTe and/or HgCdTe layers. Significant effort is needed to achieve high-quality h-BN and TMD monolayer substrates.

(b) *Nucleation control and interface engineering:* Nucleation of CdTe and HgCdTe on 2D substrates presents unique difficulties due to the absence of dangling bonds and the weak vdW interaction at the interface. These factors lead to poor wetting, low nucleation density, and the formation of randomly oriented or polycrystalline islands. To address this, various interfacial engineering strategies should be studied such as buffer layers, seeding layers, and others [60, 61].

(c) *Modified growth mechanisms on 2D substrates:* Because of the different thermal conductivity, chemical reactivity, and surface diffusion dynamics of 2D substrates growth parameters optimized for conventional 3D substrates cannot be directly applied to the growth of CdTe and HgCdTe on 2D substrates. For example, CdTe growth on h-BN was found to occur at temperatures ~100 °C lower than that on 3D substrates. This highlights the presence of a fundamentally different nucleation mechanism at the vdW interface. A systematic study of growth temperature, elemental flux ratios, and deposition rates is needed to establish an optimal growth parameter library for specific 2D substrates.

4.5 Summary

In a brief summary, 2D substrates such as graphene, mica, and others provide potential substrates for growing high-quality CdTe and HgCdTe materials without the constraint of lattice constant matching. However, most of the studies on the vdWE on these 2D substrates focus on III-N materials, and there are only a few reports on the vdWE of CdTe and HgCdTe on these 2D substrates. The material quality of the CdTe and HgCdTe layers grown on 2D substrates is not as good as expected, and defects and twin domains are often observed in these CdTe and HgCdTe layers grown. Significant effort is still needed to address those challenges as listed in the previous sections of this chapter and enhance the material quality of the CdTe and HgCdTe layers and thus their detector performance to achieve their ultimate industry applications.

References

[1] Utama M I B, Zhang Q, Zhang J, Yuan Y W, Belarre F J, Arbiol J and Xiong Q H 2013 Recent developments and future directions in the growth of nanostructures by van der Waals epitaxy *Nanoscale* **5** 3570–88

[2] Koma A 1999 Van der Waals epitaxy for highly lattice-mismatched systems *J. Cryst. Growth* **201** 236–41

[3] Pan W *et al* 2021 Large area van der Waals epitaxy of II–VI CdSe thin films for flexible optoelectronics and full-color imaging *Nano Res.* **15** 368–76

[4] Kim J, Bayram C, Park H, Cheng C W, Dimitrakopoulos C, Ott J A, Reuter K B, Bedell S W and Sadana D K 2014 Principle of direct van der Waals epitaxy of single-crystalline films on epitaxial graphene *Nat. Commun.* **5** 4836

[5] Kim Y *et al* 2017 Remote epitaxy through graphene enables two-dimensional material-based layer transfer *Nature* **544** 340–3

[6] Daudin B *et al* 2021 Growth of zinc-blende GaN on muscovite mica by molecular beam epitaxy *Nanotechnology* **32** 025601

[7] Chung K, Lee C H and Yi G C 2010 Transferable GaN layers grown on ZnO-coated graphene layers for optoelectronic devices *Science* **330** 655–7

[8] Liu Y, Xu Y, Cao B, Li Z Y, Zhao E, Yang S, Wang C H, Wang J F and Xu K 2019 Transferable GaN films on graphene/SiC by van der Waals Epitaxy for flexible devices *Phys. Status Solidi* **216** 1801027

[9] Yang Y B *et al* 2017 Surface and interface of epitaxial CdTe film on CdS buffered van der Waals mica substrate *Appl. Surf. Sci.* **413** 219–32

[10] Jung Y, Yang G, Chun S, Kim D and Kim J 2013 Growth of CdTe thin films on graphene by close-spaced sublimation method *Appl. Phys. Lett.* **103** 231910

[11] Mohanty D, Xie W Y, Wang Y P, Lu Z H, Shi J, Zhang S B, Wang G C, Lu T M and Bhat I B 2016 van der Waals epitaxy of CdTe thin film on graphene *Appl. Phys. Lett.* **109** 143109

[12] Sun X, Lu Z H, Xie W Y, Wang Y P, Shi J, Zhang S B, Washington M A and Lu T M 2017 van der Waals epitaxy of CdS thin films on single-crystalline graphene *Appl. Phys. Lett.* **110** 153104

[13] Sun X, Chen Z Z, Wang Y P, Lu Z H, Shi J, Washington M and Lu T M 2018 van der Waals epitaxial ZnTe thin film on single-crystalline graphene *J. Appl. Phys.* **123** 025303

[14] Yu J D *et al* 2019 Study on AlN buffer layer for GaN on graphene/copper sheet grown by MBE at low growth temperature *J. Alloys Compd.* **783** 633–42
[15] Novoselov K S, Geim A K, Morozov S V, Jiang D, Zhang Y, Dubonos S V, Grigorieva I V and Firsov A A 2004 Electric field effect in atomically thin carbon films *Science* **306** 666–9
[16] McCreary K M, Hanbicki A T, Robinson J T, Cobas E, Culbertson J C, Friedman A L, Jernigan G G and Jonker B T 2014 Large-area synthesis of continuous and uniform MoS_2 monolayer films on graphene *Adv. Funct. Mater.* **24** 6449–54
[17] Shi Y M *et al* 2012 van der Waals epitaxy of MoS_2 layers using graphene as growth templates *Nano Lett.* **12** 2784–91
[18] Kong W, Kum H, Bae S H, Shim J, Kim H, Kong L P, Meng Y, Wang K J, Kim C and Kim J 2019 Path towards graphene commercialization from lab to market *Nat. Nanotechnol.* **14** 927–38
[19] Kovacs A, Duchamp M, Dunin-Borkowski R E, Yakimova R, Neumann P L, Behmenburg H, Foltynski B, Giesen C, Heuken M and Pecz B 2015 Graphoepitaxy of high-quality GaN layers on graphene/6H-SiC *Adv. Mater. Interfaces* **2** 1400230
[20] Al Balushi Z Y, Miyagi T, Lin Y C, Wang K, Calderin L, Bhimanapati G, Redwing J M and Robinson J A 2015 The impact of graphene properties on GaN and AlN nucleation *Surf. Sci.* **634** 81–8
[21] Lee J H *et al* 2014 Wafer-scale growth of single-crystal monolayer graphene on reusable hydrogen-terminated germanium *Science* **344** 286–9
[22] Wang H *et al* 2016 Surface monocrystallization of copper foil for fast growth of large single-crystal graphene under free molecular flow *Adv. Mater.* **28** 8968–74
[23] Kobayashi Y, Kumakura K, Akasaka T and Makimoto T 2012 Layered boron nitride as a release layer for mechanical transfer of GaN-based devices *Nature* **484** 223–7
[24] Yu J D, Wang L, Hao Z B, Luo Y, Sun C Z, Wang J, Han Y J, Xiong B and Li H T 2020 Van der Waals Epitaxy of III-nitride semiconductors based on 2D materials for flexible applications *Adv. Mater.* **32** 1903407
[25] Yu J D *et al* 2020 Influence of nitridation on III-nitride films grown on graphene/quartz substrates by plasma-assisted molecular beam epitaxy *J. Cryst. Growth* **547** 125805
[26] Alaskar Y, Arafin S, Wickramaratne D, Zurbuchen M A, He L, McKay J, Lin Q Y, Goorsky M S, Lake R K and Wang K L 2014 Towards van der Waals epitaxial growth of GaAs on Si using a graphene buffer layer *Adv. Funct. Mater.* **24** 6629–38
[27] Chung K, Park S I, Baek H, Chung J S and Yi G C 2012 High-quality GaN films grown on chemical vapor-deposited graphene films *NPG Asia Mater.* **4** 24
[28] Yoo H, Chung K, Choi Y S, Kang C S, Oh K H, Kim M and Yi G C 2012 Microstructures of GaN thin films grown on graphene layers *Adv. Mater.* **24** 1780–0
[29] Chung K, Yoo H, Hyun J K, Oh H, Tchoe Y, Lee K, Baek H, Kim M and Yi G C 2016 Flexible GaN light-emitting diodes using GaN microdisks epitaxial laterally overgrown on graphene dots *Adv. Mater.* **28** 7688
[30] Gupta P, Rahman A A, Hatui N, Gokhale M R, Deshmukh M M and Bhattacharya A 2013 MOVPE growth of semipolar III-nitride semiconductors on CVD graphene *J. Cryst. Growth* **372** 105–8
[31] Shon J W, Ohta J, Ueno K, Kobayashi A and Fujioka H 2014 Structural properties of GaN films grown on multilayer graphene films by pulsed sputtering *Appl. Phys. Express* **7** 085502
[32] Nepal N *et al* 2013 Epitaxial growth of III-nitride/graphene heterostructures for electronic devices *Appl. Phys. Express* **6** 061003

[33] Kong W *et al* 2018 Polarity governs atomic interaction through two-dimensional materials *Nat. Mater.* **17** 999
[34] Journot T, Okuno H, Mollard N, Michon A, Dagher R, Gergaud P, Dijon J, Kolobov A V and Hyot B 2019 Remote epitaxy using graphene enables growth of stress-free GaN *Nanotechnology* **30** 505603
[35] Badokas K, Kadys A, Mickevicius J, Ignatjev I, Skapas M, Stanionyte S, Radiunas E, Juska G and Malinauskas T 2021 Remote epitaxy of GaN via graphene on GaN/sapphire templates *J. Phys. D: Appl. Phys.* **54** 205103
[36] Mohanty D *et al* 2019 Growth of epitaxial CdTe thin films on amorphous substrates using single crystal graphene buffer *Carbon* **144** 519–24
[37] Liu J L, Wang H, Li X, Chen H, Zhang Z K, Pan W W, Luo G Q, Yuan C L, Ren Y L and Lei W 2019 Ultrasensitive flexible near-infrared photodetectors based on Van der Waals BiTe nanoplates *Appl. Surf. Sci.* **484** 542–50
[38] Liu J L, Li X, Wang H, Yuan G, Suvorova A, Gain S, Ren Y L and Lei W 2020 Ultrathin high-quality SNTE nanoplates for fabricating flexible near-infrared photodetectors *ACS Appl. Mater. Interfaces* **12** 31810–22
[39] Bae S H, Kum H, Kong W, Kim Y, Choi C, Lee B, Lin P, Park Y and Kim J 2019 Integration of bulk materials with two-dimensional materials for physical coupling and applications *Nat. Mater.* **18** 550–60
[40] Ma C H *et al* 2016 Van der Waals epitaxy of functional MoO_2 film on mica for flexible electronics *Appl. Phys. Lett.* **108** 253104
[41] Ke S M, Xie J, Chen C, Lin P, Zeng X R, Shu L L, Fei L F, Wang Y, Ye M and Wang D Y 2018 van der Waals epitaxy of Al-doped ZnO film on mica as a flexible transparent heater with ultrafast thermal response *Appl. Phys. Lett.* **112** 031905
[42] Tak B R, Yang M M, Lai Y H, Chu Y H, Alexe M and Singh R 2020 Photovoltaic and flexible deep ultraviolet wavelength detector based on novel beta-Ga_2O_3/muscovite hetero-epitaxy *Sci. Rep.* **10** 19196
[43] Matsuki N, Kim T-W, Ohta J and Fujioka H 2005 Heteroepitaxial growth of gallium nitride on muscovite mica plates by pulsed laser deposition *Solid State Commun.* **136** 338–41
[44] Mohanty D *et al* 2018 Metalorganic vapor phase epitaxy of large size CdTe grains on mica through chemical and van der Waals interactions *Phys. Rev. Mater.* **2** 113402
[45] Wen X, Lu Z, Sun X, Xiang Y, Chen Z, Shi J, Bhat I, Wang G-C, Washington M and Lu T-M 2020 Epitaxial CdTe thin films on mica by vapor transport deposition for flexible solar cells *ACS Appl. Energy Mater.* **3** 4589–99
[46] Mohanty D, Sun X, Lu Z H, Washington M, Wang G C, Lu T M and Bhat I B 2018 Analyses of orientational superlattice domains in epitaxial ZnTe thin films grown on graphene and mica *J. Appl. Phys.* **124** 175301
[47] Lian Q *et al* 2019 Ultrahigh-detectivity photodetectors with Van der Waals epitaxial cdte single-crystalline films *Small* **15** 1900236
[48] Wang S F, Fong W K, Wang W and Surya C 2014 Growth of highly textured SnS on mica using an SnSe buffer layer *Thin Solid Films* **564** 206–12
[49] Lu Q J, Gao M, Lu C, Pan T S, Cheng T D, Long F and Lin Y 2021 Role of seed layer in van der Waals growth of vanadium dioxide film on mica prepared by chemical solution deposition *J. Sol-Gel Sci. Technol.* **98** 24–30
[50] Pan W W, Zhang Z K, Gu R J, Ma S, Faraone L and Lei W 2023 Van der Waals epitaxy of HgCdTe thin films for flexible infrared optoelectronics *Adv. Mater. Interfaces* **10** 2201932

[51] Ma S, Pan W W, Sun X, Zhang Z K, Gu R J, Faraone L and Lei W 2024 Growth of HgCdTe on Van Der Waals mica substrates via molecular beam epitaxy *Molecules* **29** 3947
[52] Wang J, Ma F, Liang W, Wang R and Sun M 2017 Optical, photonic and optoelectronic properties of graphene, h-BN and their hybrid materials *Nanophotonics* **6** 943–76
[53] Szczerba A K, Kucharek J, Pawlowski J, Taniguchi T, Watanabe K and Pacuski W 2023 Molecular beam epitaxy growth of cadmium telluride structures on hexagonal boron nitride *ACS Omega* **8** 44745–50
[54] Ayari T, Sundaram S, Li X, El Gmili Y, Voss P L, Salvestrini J P and Ougazzaden A 2016 Wafer-scale controlled exfoliation of metal organic vapor phase epitaxy grown InGaN/GaN multi quantum well structures using low-tack two-dimensional layered h-BN *Appl. Phys. Lett.* **108** 171106
[55] Sundaram S *et al* 2019 Large-area van der Waals epitaxial growth of vertical III-nitride nanodevice structures on layered boron nitride *Adv. Mater. Interfaces* **6** 1900207
[56] Kobayashi Y, Kumakura K, Akasaka T, Yamamoto H and Makimoto T 2014 Layered boron nitride as a release layer for mechanical transfer of GaN-based devices *Nature* **484** 223–7
[57] Park J H, Yang X, Lee J Y, Park M D, Bae S Y, Pristovsek M, Amano H and Lee D S 2021 The stability of graphene and boron nitride for III-nitride epitaxy and post-growth exfoliation *Chem. Sci.* **12** 7713–9
[58] Xu L Y, Xu Y, Qu Y P, Cao B, Wang C H and Xu K 2023 Growth mechanism of exfoliable GaN by van der Waals epitaxy on wrinkled hexagonal boron nitride *Cryst. Growth Des.* **23** 2196–202
[59] Wang L L *et al* 2024 Wafer-scale transferrable GaN enabled by hexagonal boron nitride for flexible light-emitting diode *Small* **20** 2306132
[60] Loher T, Tomm Y, Pettenkofer C, Giersig M and Jaegermann W 1995 Epitaxial-films of the 3d semiconductor Cds on the 2d layered substrate Mx(2) prepared by Van-Der-Waals epitaxy *J. Cryst. Growth* **146** 408–13
[61] Loher T, Tomm Y, Klein A, Su D, Pettenkofer C and Jaegermann W 1996 Highly oriented layers of the three-dimensional semiconductor CdTe on the two-dimensional layered semiconductors $MoTe_2$ and WSe_2 *J. Appl. Phys.* **80** 5718–22
[62] Schlaf R, Tiefenbacher S, Lang O, Pettenkofer C and Jaegermann W 1994 Van-Der-Waals epitaxy of thin inse films on $MoTe_2$ *Surf. Sci.* **303** L343–7
[63] Gupta P *et al* 2016 Layered transition metal dichalcogenides: promising near-lattice-matched substrates for GaN growth *Sci. Rep.* **6** 23708
[64] Yin Y, Ren F, Wang Y, Liu Z, Ao J, Liang M, Wei T, Yuan G, Ou H and Yan J 2018 Direct van der Waals epitaxy of crack-free AlN thin film on epitaxial WS_2 *Materials* **11** 2464

IOP Publishing

Wen Lei

Chapter 5

HgCdTe infrared detectors based on lattice-mismatched epitaxial growth

As discussed in chapters 3 and 4, significant advancement has been achieved in the MBE growth of CdTe and HgCdTe layers on lattice-mismatched alternative substrates such as Si, Ge, GaAs, and GaSb. Based on these achievements in the growth of high-quality CdTe and HgCdTe materials, this chapter will focus on the review and discussion of their applications in IR detectors and FPAs including SWIR, MWIR, and LWIR ones.

5.1 Short-wave infrared HgCdTe detectors on lattice-mismatched substrates

5.1.1 Short-wave infrared HgCdTe detectors on Si substrates

The development of SWIR HgCdTe on Si substrates began at Raytheon, USA in the early 21st century for astronomy applications. In 2002, Love *et al* and Varesi *et al* from Raytheon reported their MBE growth and processing capabilities for double-layer heterojunction (DLHJ) HgCdTe structures on 4-inch Si (211) B substrates using CdTe and ZnTe buffer layers [1, 2]. The use of Si substrates eliminates the thermal expansion mismatch between the FPAs and the ROIC, which typically constrains the maximum FPA size on other substrates. The first 1024 × 1024 SWIR p-on-n HgCdTe/Si FPAs, featuring 27 μm unit cells, were successfully fabricated for astronomical applications. These arrays exhibited cutoff wavelengths between 2.5 and 3.2 μm at 78 K, enabling observations across the J, H, and K spectral bands. The arrays demonstrated a mean responsivity of 0.164 μV/photon with a sigma-to-mean ratio of 10.9%, pixel operability of 99.12%, and an average dark current of

doi:10.1088/978-0-7503-3443-3ch5

© IOP Publishing Ltd 2025. All rights, including for text and data mining (TDM), artificial intelligence (AI) training, and similar technologies, are reserved.

approximately 0.26 $e^- s^{-1}$ at 80 K, measured on a 7.1 cm^2 device area. These low dark current levels and high operability clearly met the stringent requirements for astronomical imaging.

In 2008, Reddy *et al* from Raytheon advanced the SWIR HgCdTe/Si technology by increasing the array size to 2048 × 2048 pixels and reducing the unit cell pitch to 15 μm on 4-inch Si substrates [3]. Their results demonstrated exceptionally uniform composition across the wafers, independent of the CdTe mole fraction or substrate size. The wafers also exhibited uniformly low surface defect densities, typically below 10 cm^{-2}, and not exceeding 25 cm^{-2} even in the worst cases. Although detailed device performance was not reported, Raytheon has indicated plans to develop larger arrays with dimensions up to 4k × 4k pixels in future work.

Apart from Raytheon, other institutions also studied the SWIR HgCdTe detectors on Si substrates. In 2014, Bommena *et al* from EPIR, USA demonstrated high-performance SWIR HgCdTe detectors with a device structure of double layer planar heterostructure (DLPH) grown on 3-inch CdTe-buffered Si substrates using MBE [4]. The grown layers exhibited minority carrier lifetimes in the order of 3 μs at room temperature. Electrical characterization of the devices revealed diffusion-limited performance at the operating bias, with a low dark current density on the order of 10^{-6} A cm^{-2} at 200 K and 10^{-3} A cm^{-2} at 300 K. Optical measurements showed a quantum efficiency exceeding 70% at 2.0 μm, even without an antireflective coating, which is comparable to the state-of-the-art SWIR HgCdTe detectors grown on CdZnTe substrates. FPAs with 320 × 256 resolution and 30 μm pitch were fabricated with a cutoff wavelength of 2.65 μm. These arrays demonstrated a mean dark current of 30 pA/pixel at 200 K and a pixel operability of 99.5%. The material and detector performance exceeded those of III–V compound semiconductors, which are commonly employed for SWIR detection.

In 2016, Park *et al* from EPIR reported extended short-wavelength infrared (eSWIR) FPAs fabricated from MBE-grown n-type HgCdTe on 3-inch Si substrates, with a cutoff wavelength of 2.68 μm at 77 K [5]. The grown material exhibited high uniformity in both composition and thickness across the 3-inch wafer, along with low surface defect densities (9.56×10^1 cm^{-2} for voids and 1.67×10^3 cm^{-2} for micro-defects). Using this material, 320 × 256 format FPAs with 30 μm pitch were fabricated, employing a planar p-on-n device architecture with arsenic implantation to achieve p-type doping. Figure 5.1(a) shows the device structure of the eSWIR p-on-n DLPH. The test devices exhibited uniform dark current density across the temperature range from 190 K to room temperature. Hybridized FPAs demonstrated high-quality eSWIR imaging, with a pixel operability of 99.27%, and a median dark current density of 2.63×10^{-7} A cm^{-2} and a standard deviation of 1.67×10^{-7} A cm^{-2} at 193 K. Excellent image contrast and quality were observed at 193 K and 85 K, as shown in figures 5.1(b) and (c). Later that year, the FPA format size was successfully scaled up to 640 × 512 pixels, maintaining similar device performance [6].

In 2017, Hu *et al* from Shanghai Institute of Technical Physics (SITP) demonstrated a 512 × 512-pixel SWIR HgCdTe diode array with 30 μm pitch on a 3-inch Si

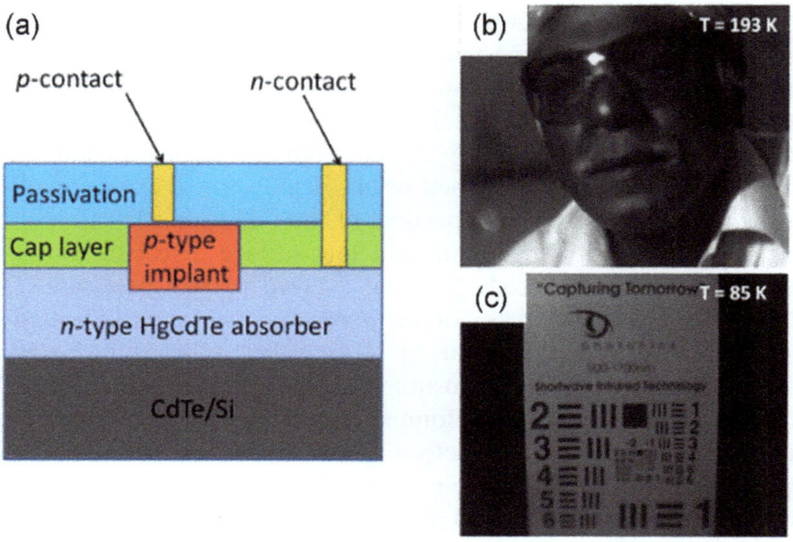

Figure 5.1. (a) Schematic p-on-n DLPH device structure, imaging results of the 320 × 256 SWIR HgCdTe FPAs grown on Si at (b) 193 K and (c) 85 K. Panels (a)–(c) were reproduced from [5] with permission from Springer Nature.

substrate for hyperspectral imaging [7]. By precisely stitching four 512 × 512 HgCdTe/Si FPAs, a mosaic array with a 2000 × 512 format was realized. Characterization of the mosaic array revealed a dark current density on the order of 1.67×10^{-10} A cm^{-2}, a quantum efficiency exceeding 70%, and an operability of 99.5% at an operating temperature of approximately 110 K. Under illumination with 5×10^4 photons/pixel, the FPA achieved an SNR of 120 which was excellent for imaging.

5.1.2 Short-wave infrared HgCdTe detectors on Ge substrates

Although there were a number of institutions around the world working on the MBE growth of CdTe and HgCdTe materials on Ge substrates, there were not many reports on the SWIR detector applications. Instead, the study of IR detector applications of HgCdTe grown on Ge substrates mainly focuses on MWIR and LWIR detectors which will be discussed in the next two sections of this chapter. This might be due the fact that SWIR detectors are less sensitive to defects in comparison to MWIR and LWIR detectors due to their wider energy bandgap. Thus, if HgCdTe grown on Ge substrates can be used for making high-performance MWIR and LWIR detectors, they can also be used for making high-performance SWIR detectors. The researchers from Kunming Institute of Physics, China, reported their study on MBE growth of HgCdTe materials on Ge substrates and their applications in SWIR, MWIR, and SWIR/MWIR dual band detectors [8]. In their work, the SWIR HgCdTe layers grown on 3-inch Ge substrates presented an XRD FWHM of

86 arcsec, an EDP of 2.9×10^6 cm^{-2}, and surface defect density of 200 cm^{-2}. Based on the high-quality HgCdTe materials obtained, they fabricated a SWIR HgCdTe FPA with a format of 320 × 256, and the SWIR HgCdTe FPA showed a peak detectivity of 2.6×10^{12} cm Hz$^{1/2}$ W^{-1}, a NEDT of 20.3 mK, and a pixel operability of >99.92%. The performance of these SWIR HgCdTe detectors and FPAs is comparable to those grown on lattice-matched CdZnTe substrates.

5.1.3 Short-wave infrared HgCdTe detectors on GaAs substrates

Similar to the case of Ge alternative substrates, there were limited reports on HgCdTe SWIR detectors grown on GaAs substrates. In 2002, Kim *et al* from the Korea Institute of Science and Technology reported the MOCVD growth of SWIR HgCdTe on GaAs (001) substrates using a CdTe buffer layer [9]. The resulting n-on-p HgCdTe/GaAs structure was fabricated into a backside-illuminated, mesa-etched photodiode array consisting of 32 rectangular diodes, each with a junction area of 110 × 110 μm^2. The zero-bias dynamic resistance–area product (R_0A) values were 5×10^5 Ω cm^2 at 200 K and 3×10^2 Ω cm^2 at 300 K, which is comparable to planar photodiodes fabricated using conventional ion implantation in MBE-grown HgCdTe on CdZnTe substrates [10]. The relative spectral response of the devices at 300 K showed a cutoff wavelength of approximately 2.5 μm. These results show that the MOCVD growth of HgCdTe on GaAs could be used to fabricate large-area FPAs in the SWIR spectral range.

In 2005, Ye *et al* from SITP reported the fabrication of MBE-grown HgCdTe two-color (SWIR/MWIR) FPAs based on a four-layer p–P–P–N heterostructure grown on GaAs (211)B substrates [11]. The structure included a p-type absorption layer on top of a thin p-type potential barrier layer, and the SWIR p-on-n homojunction photodiode was formed *in situ* during MBE growth via indium doping. Preliminary 256 × 1 linear SWIR/MWIR two-color HgCdTe FPAs were fabricated using mesa isolation, sidewall passivation, and contact metallization. At 78 K, the average R_0A values for the SWIR and MWIR photodiodes were 3.85×10^5 Ω cm^2 and 3.02×10^5 Ω cm^2, respectively. The corresponding average peak detectivities were 1.57×10^{11} cm Hz$^{1/2}$ W^{-1} for SWIR and 5.63×10^{10} cm Hz$^{1/2}$ W^{-1} for MWIR. The cutoff wavelengths were measured at 3.04 μm for the SWIR photodiode and 5.74 μm for the MWIR photodiode. Further improvements in quantum efficiency and minority carrier lifetime, particularly in the shorter wavelength regime, could be achieved by employing a SWIR p-on-n heterostructure rather than a homojunction design.

In a short summary, the development of SWIR HgCdTe detectors primarily focused on Si substrates, and very limited reports were available regarding SWIR HgCdTe detectors on Ge and GaAs substrates. But overall, SWIR HgCdTe detectors on Si, Ge, and GaAs substrates presented a high device performance comparable to those on CdZnTe substrates. To date, the largest reported array format for SWIR HgCdTe FPAs is 2048 × 2048 with a 15 μm pixel pitch, fabricated on 4-inch Si substrates. Although a larger array format such as 4k × 4k is a key goal for future IR detector development, achieving this remains technically challenging and is still under active development. Moreover, most high-performance SWIR FPAs are used for astronomy and scientific imaging applications, which demand ultra-low dark current and excellent uniformity, and thus set stringent requirements for large-format SWIR HgCdTe detectors and FPAs on Si substrates.

5.2 Mid-wave infrared HgCdTe detectors on lattice-mismatched substrates

5.2.1 Mid-wave infrared HgCdTe detectors on Si substrates

As discussed in chapter 3, significant effort has been devoted to the MBE growth of CdTe and HgCdTe layers on Si alternative substrates, especially in the 1990s, leading to significant advancement in their IR detector applications. In 1998, De Lyon *et al* from Hughes Research Lab, USA, demonstrated the first high-performance MWIR HgCdTe detectors grown on Si substrates via MBE [12]. In their study, an HgCdTe p^+-on-n double-layer heterojunction structure was grown on 3-inch Si (211) substrates with a CdTe buffer layer. The resulting HgCdTe layers exhibited high crystalline quality with XRD FWHMs in the range from 75 to 100 arcsec and EPDs in the range from 2×10^7 cm^{-2} to 5×10^7 cm^{-2}. A small-format FPA was fabricated with an external quantum efficiency of ~70%. Although slightly larger $1/f$ noise was observed due to the higher dislocation density in comparison to HgCdTe grown on CdZnTe substrates, the overall device performance remained comparable to that grown on CdZnTe. Specifically, R_0A products reached 10^6–10^7 Ω cm^2 for a 3.6 μm cutoff at 125 K, and 10^4–10^5 Ω cm^2 for a 4.7 μm cutoff at the same temperature, matching the performance of HgCdTe detectors grown on lattice-matched CdZnTe [12].

In 1999, the joint work between the U.S. Army Research Laboratory and CEA LETI, France, made further advancements in this area [13]. In this work [13], Dhar *et al* studied the critical role of initial ZnTe nucleation layer and CdTe buffer layer in improving both the crystalline quality and surface morphology of the overgrown HgCdTe films on 2-inch Si substrates. The HgCdTe layers exhibited an EPD around the order of 2×10^6 cm^{-2}, indicating substantial reduction in threading dislocations compared to direct growth. Based on the HgCdTe layers grown, n-on-p MWIR photodiodes were fabricated with the format of 34-element linear diode arrays. The devices showed excellent performance such as shunt impedance of 3.04 GΩ at a peak wavelength of 4.79 μm at 77 K and uniform quantum efficiency of 60%, demonstrating the importance of buffer layer strategy and the potential of Si substrates for making high-performance MWIR HgCdTe detectors.

Raytheon, a global leading manufacturer of HgCdTe IR detectors, also significantly advanced the development of MBE-grown MWIR HgCdTe detectors on Si substrates in the early 21st century. Motivated by the demand for larger array formats and smaller pixel sizes, Raytheon initially focused on MBE growth of high-quality HgCdTe on 4-inch Si wafers. In 2001, Maranowski *et al* from Raytheon successfully demonstrated the MBE growth of p-on-n DLHJ HgCdTe

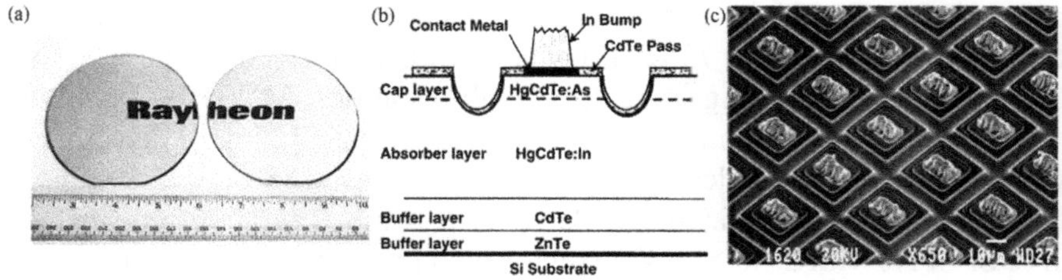

Figure 5.2. (a) Photo of two high-quality 4-inch HgCdTe-on-Si wafers grown by MBE. (b) Schematic cross-section view of the HgCdTe/Si DLHJ structure. (c) SEM image of a 128 ×128 format HgCdTe FPA with 40 μm pitch. Reproduced from [15] with permission from Springer Nature.

structures on 4-inch Si substrates for MWIR detector applications [14] (figure 5.2 (a)). The HgCdTe epitaxial films exhibited excellent structural quality with XRD FWHMs ranging from 80 to 100 arcsec, EPDs ranging from 1×10^6 cm^{-2} to 7×10^6 cm^{-2}, and superior compositional and thickness uniformity across large-diameter wafers. Also in 2001, Varesi *et al* from Raytheon fabricated high-performance MWIR detectors (5.7 μm cut-off) based on the DLHJ HgCdTe/Si wafers [15] (figure 5.2(b)). The quantum efficiency of these MWIR detectors reached 69% at a wavelength of 4.0 μm and an operating temperature of 140 K. Note that the 69% quantum efficiency is slightly lower than the typical maximum quantum efficiency of 79% observed for CdZnTe-based MWIR HgCdTe detectors [15], which can be mainly attributed to the higher reflectivity of Si surface, and can be mitigated through the application of anti-reflection coatings. Importantly, the R_0A values of the HgCdTe/Si detectors (5.3×10^3 Ω cm^2) approached the theoretical radiative limit and outperformed those of the counterpart LPE-grown HgCdTe detectors (3×10^3 Ω cm^2). Although the $1/f$ noise was approximately twice as high as that of LPE-grown HgCdTe on CdZnTe, it remained sufficiently low to avoid any significant degradation in the FPA performance. A 128 × 128 FPA with a 40 μm unit cell (figure 5.2(c)) was also fabricated with MWIR sensitivity comparable to that of mature InSb technology, achieving pixel operability greater than 99%. They also demonstrated a 640 × 480 array based on HgCdTe materials grown on Si which however only showed a pixel operability of 98%.

Further advancement was reported by Vilela *et al* from Raytheon in 2005, who further expanded the FPA format to 2560 × 512 pixels using p-on-n DLHJ MWIR HgCdTe structures grown on 4-inch Si substrates via MBE [16]. The HgCdTe material demonstrated excellent structural and electrical properties, with an XRD FWHM as low as 64 arcsec and a typical EPD around mid-10^6 cm^{-2}. Cutoff wavelength uniformity across the entire wafer was maintained within 5%, and the density of surface macro-defects was reported to be below 1000 cm^{-2}, indicating high crystalline quality suitable for large-format integration. Device-level performance, including R_0A values ranging from mid-10^5 to high-10^6 Ω cm^2, a spectral response with a cutoff wavelength around 5.2 μm at 78 K, strong reverse breakdown in current–voltage (I–V) characteristics, and quantum efficiency of 60%–70% for the 2560 × 512 FPAs, demonstrated the increasing maturity of HgCdTe/Si detector technology. These 2560 × 512 FPAs presented a response operability well over 99% and NEDT below 20 mK at 78 K [17]. These results underscored the viability of Si as a substrate for fabricating large-area, high-resolution MWIR imaging systems with performances approaching that of traditional CdZnTe-based detectors.

By 2011, Raytheon extended the MBE growth of HgCdTe on Si to 6-inch Si (211) substrates for imaging applications. Bangs *et al* from Raytheon reported MWIR HgCdTe/Si FPAs with 2048 × 2048 pixel format (15 μm pitch), showing outstanding operability (99.9%) and stable quantum efficiency of approximately 60%, varying by less than 4% over a temperature range from 180 K to 85 K. A NEDT of 23.1 mK was achieved [18]. Moreover, HgCdTe devices fabricated on 6-inch Si substrates demonstrated comparable FPA performance to those on 4-inch HgCdTe/Si substrates, as reported by Raytheon [15], and consistently approached the performance of HgCdTe/CdZnTe devices.

In 2016, Raytheon also reported the state-of-the-art 4k × 4k, 20 μm pitch MWIR HgCdTe detector array grown on Si substrate by MBE [19], as shown in figure 5.3. This array represented the largest infrared FPA fabricated to date using HgCdTe/Si technology [19], demonstrating Raytheon's strong capability in fabricating large-format, small-pixel arrays.

SITP also reported its work on MWIR HgCdTe detectors grown on Si substrates. In the SITP study [20], HgCdTe layers of different compositions were subsequently grown on the CdTe buffer layers via a graded-composition region to reduce the lattice mismatch (0.2%) between CdTe and HgCdTe. The HgCdTe layers grown presented an XRD FWHM in a range of 55–75 arcsec, an EPD value in a range of 1×10^6 cm^{-2} to 5×10^6 cm^{-2}, and surface defect density of <500 cm^{-2}. The MWIR

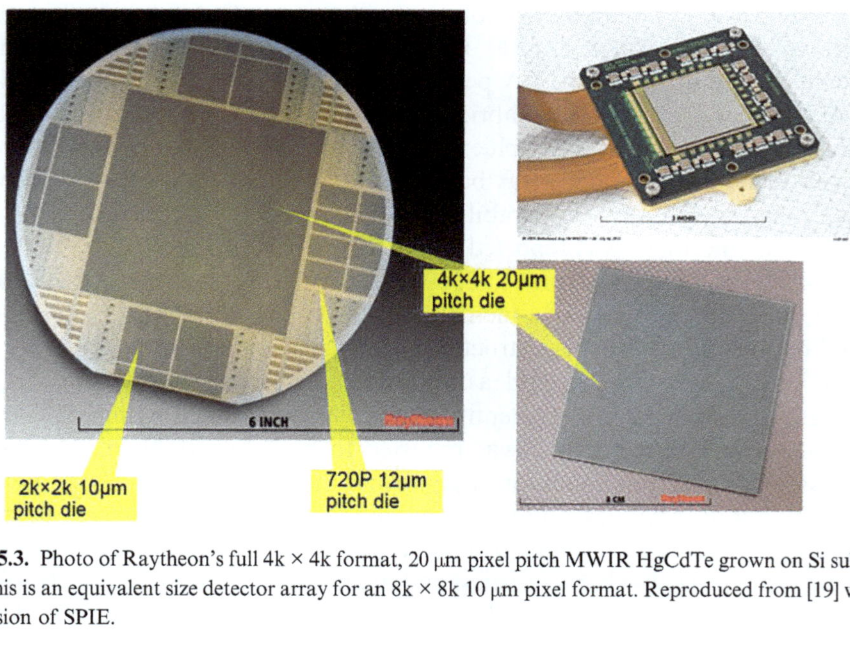

Figure 5.3. Photo of Raytheon's full 4k × 4k format, 20 μm pixel pitch MWIR HgCdTe grown on Si substrate. Note this is an equivalent size detector array for an 8k × 8k 10 μm pixel format. Reproduced from [19] with the permission of SPIE.

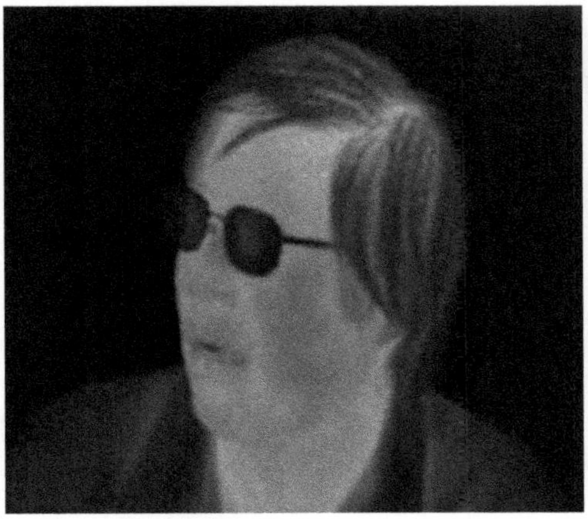

Figure 5.4. Imaging result with a SITP's MWIR HgCdTe FPA (320 × 240) grown on Si substrate. Reproduced from [20] with permission from Springer Nature.

FPAs were fabricated with planar n+p homojunctions formed by using B^+ implantation into p-type HgCdTe epilayers. The MWIR HgCdTe FPA grown on Si showed a detectivity D^* of 2×10^{11} cm $Hz^{1/2}$ W^{-1} and a pixel operability above 98% at 80 K. Figure 5.4 showed the imaging result with a SITP's MWIR HgCdTe FPA grown on Si substrate. The results are comparable with that of FPAs by HgCdTe grown on lattice-matched CdZnTe substrates, suggesting that the alternative composite substrates of CdTe/Si may be capable of replacing CdZnTe for MWIR detector applications.

5.2.2 Mid-wave infrared HgCdTe detectors on Ge substrates

In parallel with the study of MWIR HgCdTe on Si substrates, significant advancement has been made in the MBE growth and detector fabrication of MWIR HgCdTe on Ge substrates. Note that the study of HgCdTe on Ge was primarily undertaken by CEA LETI and Sofradir, France. In 1998, Zanatta *et al* from LETI reported the MBE growth of HgCdTe layers on Ge (211) substrates, and studied their application in MWIR detectors [21]. The HgCdTe layers exhibited an average XRD FWHM of 105 arcsec with a best of 68 arcsec. The small MWIR detector arrays (32 pixels) fabricated showed a cutoff at 4.8 μm, a peak response (4.6 μm) of 2.2 A W^{-1}, a quantum efficiency of 60%, a peak detectivity D^* of 6×10^{11} cm Hz W^{-1} (4.6 μm), and a R_0A product of $>10^6$ Ω cm^2 at 78 K. Building upon this foundation, Ferret *et al* from LETI expanded the device scale to a 320 × 240 array with a 30 μm pixel pitch in 2000 [22]. The resultant FPAs showed a cut-off wavelength of 5.0 μm, a mean NEDT of 8.8 mK and an operability of 99.94%, demonstrating both material uniformity and high device yield.

Later, in 2006, the joint work between Sofradir and LETI scaled up wafer size by growing MWIR HgCdTe on 3- and 4-inch Ge substrates (see figure 5.5(a)) [23]. MBE-grown HgCdTe on 4-inch Ge substrate has achieved material quality comparable to that grown with LPE. Reported EPD values reached the low

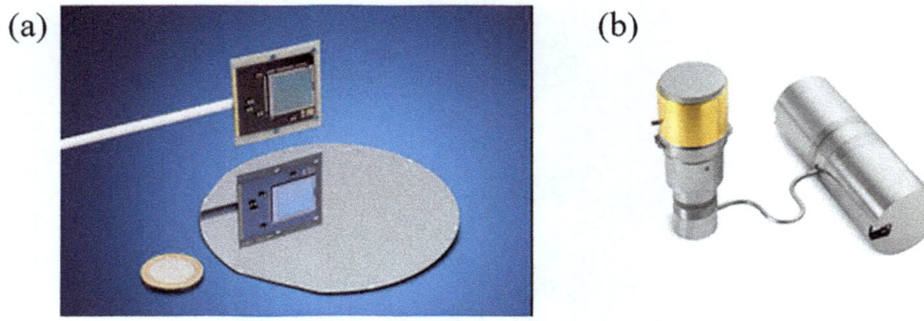

Figure 5.5. (a) Photo of a 4-inch HgCdTe/Ge wafer. (b) FPA module of Jupiter 1280 × 1024 MWIR HgCdTe detector FPA. (Reproduced from [23] with the permission of SPIE.)

10^6 cm^{-2} range, and XRD FWHM values was between 90 and 130 arcsec, indicating high crystalline quality [23]. One of the most advanced MWIR HgCdTe detector systems developed on Ge is Sofradir's 'Jupiter' FPA product (see figure 5.5(b)) [23]. The 'Jupiter' FPAs have a large 1280 × 1024 format with 15 μm pitch fabricated on 4-inch Ge substrates, which is a typical product of MWIR detectors sensing in the 3–5 μm wavelength range. The FPAs present a cut-off wavelength of 5.5 μm, an average NEDT of 19 mK, a quantum efficiency exceeding 70%, a fill factor above 80%, and an operability greater than 99.8% at 80 K. The demonstrated performance showed the technological maturity and practical viability of HgCdTe/Ge FPAs for high-resolution IR imaging.

5.2.3 Mid-wave infrared HgCdTe detectors on GaAs substrates

As discussed in chapter 3, GaAs is another intensively studied alternative substrate for growing HgCdTe to achieve HgCdTe IR detectors with features of lower cost and larger array format size. Note that as discussed in chapter 3, HgCdTe layers grown on GaAs substrates were demonstrated to have the lowest EPD (2.3×10^5 cm^{-2}) after applying cyclic thermal annealing in comparison to those grown on Si and Ge substrates. So, HgCdTe IR detectors grown on GaAs should potentially have better device performance than those grown on Si and Ge.

Dvoretsky *et al* from Institute of Semiconductor Physics, Russia, reported the MBE growth of HgCdTe on 2-inch GaAs substrates in 2005 [24]. The MWIR HgCdTe layers grown presented a high electron mobility of 5×10^4 cm^2 V^{-1} s^{-1} and a long minority carrier lifetime around 10 μs. The single MWIR HgCdTe detector fabricated showed a peak specific detectivity D^* of 8.5×10^{10} cm $Hz^{1/2}$ W^{-1} at a cutoff wavelength of 5.4 μm when operated at 78 K, and 2.3×10^{10} cm $Hz^{1/2}$ W^{-1} at a cutoff wavelength of 5 μm at 200 K, highlighting their potential for both cryogenic and moderately elevated temperature operation. 320 × 256 MWIR FPAs with 40 μm pitch were also fabricated which presented a cutoff wavelength of 5.5 μm, a NEDT of 23 mK, and a pixel operability greater than 98% at 78 K. These results confirm the viability of GaAs-based platforms for high-resolution, high-sensitivity MWIR imaging applications.

SITP also reported its MWIR HgCdTe detectors grown on GaAs substrates in 2008. In the SITP study [20], similarly HgCdTe of different compositions was subsequently grown on the CdTe buffer layer via a graded-composition region to reduce the lattice mismatch (0.2%) between CdTe and HgCdTe. The HgCdTe layers grown presented an XRD FWHM in a range of 55–75 arcsec, an EPD value in a range from 1×10^6 cm^{-2} to 5×10^6 cm^{-2}, and a surface defect density of <300 cm^{-2}. The MWIR FPAs were fabricated with planar n+p homojunctions formed by using B^+ implantation into p-type HgCdTe epilayers. The MWIR HgCdTe FPA grown on GaAs showed a detectivity D^* of 2×10^{11} cm Hz$^{1/2}$ W^{-1} and a pixel operability above 98% at 80 K. Figure 5.6 shows the imaging results using the SITP's MWIR HgCdTe FPAs grown on GaAs substrates. The results are comparable with that of FPAs grown on lattice matched CdZnTe substrates, suggesting that the alternative composite substrates of CdTe/GaAs may also be capable of replacing CdZnTe for MWIR detector applications.

Further advancement was reported in 2015 by Wenisch *et al* from AIM Infrarot-Moduleg GmbH, Germany, who demonstrated MBE growth of HgCdTe on 4-inch GaAs (211)B substrates [25]. A 640 × 512 MWIR HgCdTe FPA with 15 μm pixel pitch was fabricated using a planar n-on-p photodiode manufacturing process. The FPA exhibited excellent performance with an operability of 99.71% and NEDT of 20.9 mK at an operating temperature of 120 K, and a low dark current density of approximately 3 fA μm^{-2}. At AIM, they also aimed to develop MWIR HgCdTe FPAs on GaAs with an even larger array of format sizes such as a 1280 × 1024 format.

In short summary, significant advancement has been made in the MBE growth of MWIR HgCdTe materials on Si, Ge, and GaAs, and high-performance MWIR HgCdTe FPAs have been demonstrated. Note that although CdTe layers grown on

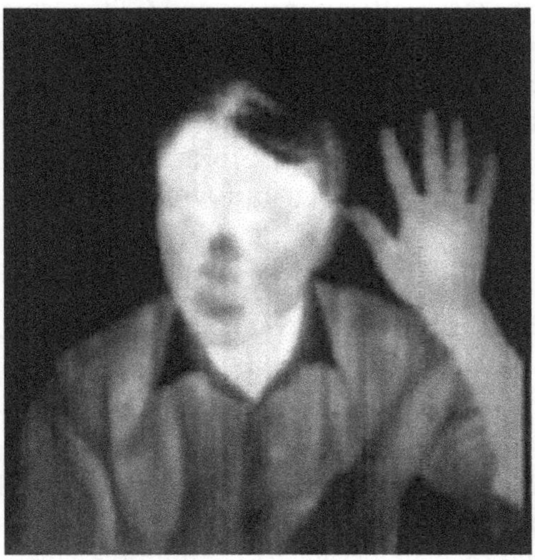

Figure 5.6. Imaging result with a SITP's MWIR HgCdTe FPA (128 × 128 format) grown on GaAs substrate. (Reproduced from [20] with permission from Springer Nature.)

GaSb presented a very low EPD (~1.5×10^5 cm^{-2}), no detector and/or FPA applications have been reported yet. This is mainly because so far, no industry was involved in the development of GaSb alternative substrate technology, which slowed down the whole process. The industry has already invested heavily in Si, Ge, and GaAs alternative substrate technologies and would not easily switch to another new substrate technology. From what has been demonstrated so far, HgCdTe layers on alternative substrates could be grown up to 6-inch wafers, and the MWIR HgCdTe detectors grown on Si, Ge, and GaAs substrates showed device performance comparable to that on lattice-matched CdZnTe substrates, which led to industry products such as Jupiter from Sofradir, France. The fabrication of HgCdTe on these alternative substrates does facilitate the development of HgCdTe detectors with the features of lower cost and larger array format size. However, it should be noted that the threading dislocation density (mid-10^6 cm^{-2} to low-10^7 cm^{-2} on average) in HgCdTe layers grown on these alternative substrates is much higher than that (low-10^4 cm^{-2} to low-10^5 cm^{-2} on average) grown on lattice-matched CdZnTe substrates. This indicates that the performance up-limit of these MWIR HgCdTe detectors on alternative substrates will be lower than that on CdZnTe substrates although the current MWIR detector performance might be sufficient for applications. So, further effort is still needed to further drive down the dislocation density in HgCdTe layers grown on alternative substrates to a level comparable to that on CdZnTe substrates.

5.3 Long-wave infrared HgCdTe detectors on lattice-mismatched substrates

Compared with SWIR and MWIR detector applications, LWIR detectors raised much more stringent requirements on the material quality of HgCdTe epilayers. However, because of the large lattice constant and CTE mismatch between HgCdTe and alternative substrates (as discussed in chapters 1 and 3), it is inevitable that a high density of threading dislocations are generated when epitaxially growing HgCdTe layers on lattice-mismatched alternative substrates [26]. Therefore, it is much more challenging to achieve high-performance LWIR HgCdTe detectors on alternative substrates. Despite this, some progress has been made in this area as detailed below.

5.3.1 Long-wave infrared HgCdTe detectors on Si substrates

So far, a number of institutions have reported their work on the LWIR detector application of HgCdTe materials grown on Si substrates. Teledyne group from the United States grew their HgCdTe layers on Si (211) substrate with ZnTe/CdTe buffer layers to mitigate the negative effects of large lattice and CTE mismatch [27]. A double-layer planar heterostructure (DLPH) structure was used for fabricating the LWIR HgCdTe detectors. This design had a multiple layer structure including a metal contact, an n-type LWIR HgCdTe absorber layer (~10 μm thick), an MBE-grown CdTe passivation layer, an MWIR HgCdTe cap, and CdTe/ZnTe buffer layers. N-type doping was achieved by indium doping during MBE growth, while p-type doping was achieved with ion implantation [27].

Figure 5.7(a) shows the quantum efficiency (QE) response of the LWIR HgCdTe detectors at the temperature of 78 K, alongside the theoretical model fit based on an assumed minority carrier diffusion length of 8 μm. This extracted diffusion length is roughly half of the value typically observed in the LWIR HgCdTe layers grown on lattice-matched CdZnTe substrates [27]. The reduced minority carrier lifetime and thus diffusion length could be attributed to the higher dislocation density in the HgCdTe epitaxial layers, which exceeded 10^6 cm^{-2}, and adversely impacted carrier transport properties. LWIR FPAs with a 640 × 480 format and 20 μm pitch were also fabricated and tested to evaluate the feasibility of HgCdTe layers grown on Si for applications in LWIR detectors. The LWIR HgCdTe-on-Si FPAs showed a pixel operability over 98%, with both mean and median NEDTs below 25 mK at an operating temperature of 78 K and under high background flux conditions. Figure 5.7(b) shows the 78 K NEDT histogram of a 640 × 480 format, 20 μm pitch LWIR HgCdTe-on-Si FPA [27]. Despite these encouraging results, the performance of LWIR HgCdTe detectors grown on Si substrates still fell short of that of grown on lattice-matched CdZnTe substrates, particularly in terms of pixel uniformity and overall detector performance. Note that the quantum efficiency of LWIR HgCdTe-on-Si detectors was also lower than that of the counterpart devices grown on CdZnTe [28]. These limitations indicated that further improvement in material quality, especially threading dislocation density and resultant electrical properties (minority carrier lifetime, carrier mobility, etc) are highly required before achieving LWIR HgCdTe-on-Si detectors with performance able to compete with those grown on CdZnTe substrates. This is mainly because LWIR detectors are more sensitive to the dislocations in comparison to SWIR and MWIR detectors due to their much narrower energy bandgaps.

Raytheon also reported its work on LWIR HgCdTe detectors grown on Si substrates [17]. The Raytheon researchers grew HgCdTe epilayers on 4- and 6-inch Si wafers with high uniformity (cutoff and surface defects), and used a DLHJ

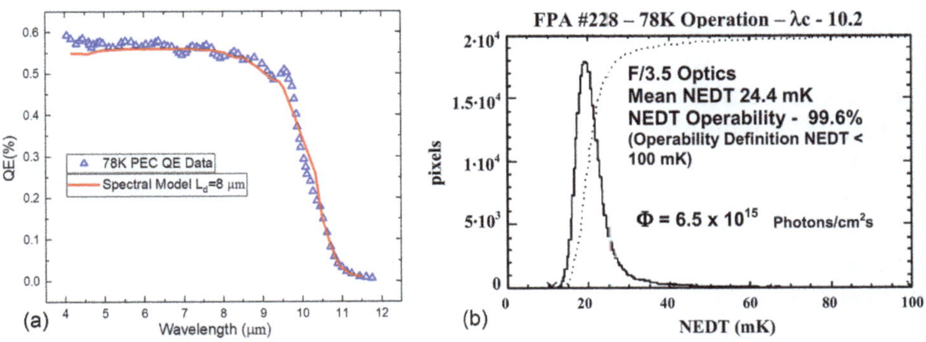

Figure 5.7. (a) Quantum efficiency (QE) spectral response of the DLPH diode fabricated at Teledyne (blue triangles), and spectral model fit assuming minority carrier diffusion length of 8 μm. (b) NEDT histogram of the 640 × 480 LWIR HgCdTe-on-Si FPA. Reproduced from [27] with permission from Springer Nature.

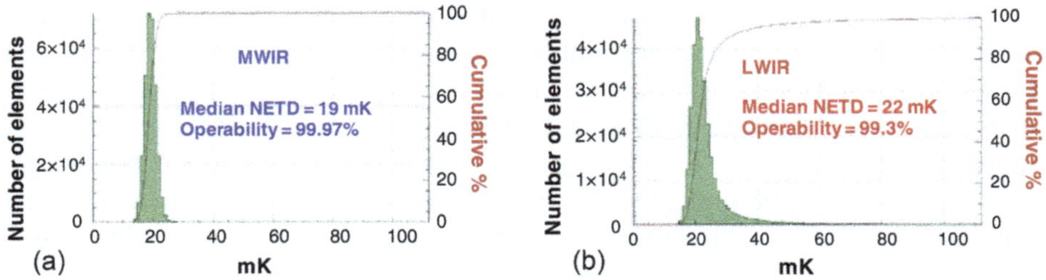

Figure 5.8. NEDT histograms of both the (a) MWIR and (b) LWIR bands of Raytheon's two-color layer on Si at 78 K. Reproduced from [17] with permission from Springer Nature.

detector structure for fabricating HgCdTe devices on Si. Typically, the surface defect density was in the range of several 10 to several 100 cm^{-2}, and the EPD was around mid-10^6 cm^{-2}. Two-color MWIR/LWIR 640 × 480 FPAs were fabricated based on the HgCdTe materials grown on Si. Figure 5.8 showed the NEDT histograms of both the MWIR and LWIR bands of two-color layer on Si at 78 K. The NEDT operability of the MWIR band shows a very high value of 99.97%, and the operability drops slightly to 99.3% for the LWIR band.

Apart from Teledyne and Raytheon, SITP also reported its LWIR HgCdTe detectors grown on Si substrates [20]. In the SITP study [20], again HgCdTe of different compositions was subsequently grown on the CdTe buffer layer via a graded-composition region to reduce the lattice mismatch (0.2%) between CdTe and HgCdTe. The HgCdTe layers grown presented an XRD FWHM in a range of 55–75 arcsec, an EPD value in a range from 1×10^6 cm^{-2} to 5×10^6 cm^{-2}, and surface defect density of <500 cm^{-2}. The LWIR HgCdTe FPA was fabricated with an n-on-p mesa architecture of grown p–n heterojunctions. The LWIR HgCdTe FPA grown on Si showed a detectivity D^* of 6×10^{10} cm $Hz^{1/2}$ W^{-1} and a pixel operability above 99% at 80 K.

5.3.2 Long-wave infrared HgCdTe detectors on Ge substrates

As discussed in chapters 1 and 3, in comparison to Si substrates, Ge substrates present a smaller lattice and CTE mismatch, and in principle, should lead to HgCdTe epilayers with better material quality [26, 29]. In 2008, Raytheon researchers successfully demonstrated the first LWIR HgCdTe-on-Ge FPAs with the structure of p-on-n DLHJs [30]. The LWIR HgCdTe epilayers were grown on Ge (211) substrates with ZnTe/CdTe buffer layer by MBE technique. The EDP level of the LWIR HgCdTe epilayers was measured to be within mid-10^6 cm^{-2} range, which is at the same EPD level of LWIR HgCdTe epilayers grown on Si (211) substrates. To compare the detector performance between LWIR HgCdTe-on-Ge and LWIR HgCdTe-on-Si devices, the same p-on-n DLHJ HgCdTe detector structure was grown on both Ge and Si substrate at Raytheon. At 78 K, the NEDTs of the 256 × 256 FPAs fabricated on Ge and Si substrates were measured to be 17.2 and 14.5 mK, respectively. Corresponding pixel operability was 99.0% for Ge-based LWIR HgCdTe FPAs and 98.7% for Si-based LWIR HgCdTe FPAs

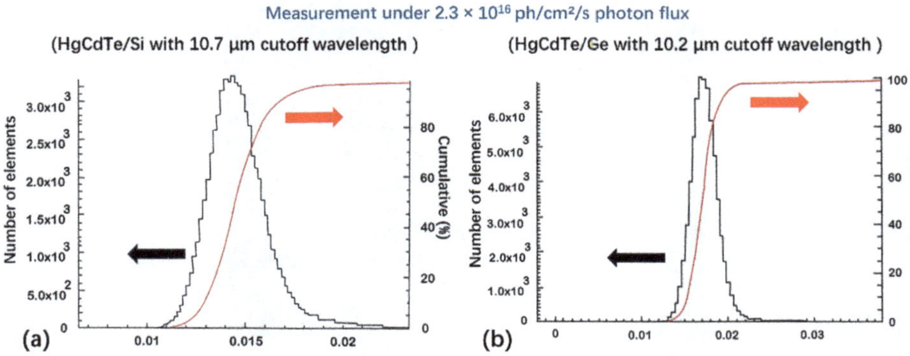

Figure 5.9. Pixel operability for the LWIR HgCdTe FPAs fabricated on (a) Si and (b) Ge substrates under the condition of a flux of 2.3×10^{16} ph cm^{-2} s^{-1}. Reproduced from [30] with the permission of Springer.

under an illumination flux of 2.3×10^{16} ph cm^{-2} s^{-1}, as illustrated in figure 5.9. It was also observed that the R_oA product of 256 × 256 LWIR HgCdTe FPAs on Ge was very similar to that grown on Si. Both types of LWIR detectors presented a quantum efficiency in the 60%–70% range without an antireflection coating. As a result, Raytheon came to the conclusion that LWIR HgCdTe devices grown on Si and Ge presented similar device performance with the same growth and device fabrication processes. This might be the reason that no more study on Ge alternative substrates was reported from Raytheon, and only that on Si alternative substrates were found from Raytheon afterwards.

5.3.3 Long-wave infrared HgCdTe detectors on GaAs substrates

Apart from Si and Ge, GaAs substrates were also studied for growing LWIR HgCdTe detectors by a number of institutions. Dvoretsky *et al* from the Institute of Semiconductor Physics, Russia, used an intermediate CdZnTe (0.05 μm ZnTe and 6.5 μm CdTe) epilayer as the buffer layer to eliminate the effect of mixed phases at the interface between II–VI material (HgCdTe) and III–V substrate (GaAs) [24]. LWIR 320 × 240 HgCdTe FPAs were fabricated based on p-type HgCdTe grown on GaAs. The average NEDT value was measured to be 19 mK for a 10.2 μm cut-off wavelength at the operating temperature of 78 K. In addition, the pixel operability was measured to be over 95%. Obviously, the device performance for LWIR HgCdTe detectors grown on GaAs is much lower than on CdZnTe substrates, especially the low pixel operability, which is not sufficient for practical applications.

The group from SITP also reported LWIR HgCdTe detectors based on HgCdTe grown on GaAs [20]. Similar to the HgCdTe layers grown on Si at SITP, HgCdTe layers grown on GaAs presented an XRD FWHM in a range of 55–75 arcsec, an EPD value in a range of 1×10^6 cm^{-2} to 5×10^6 cm^{-2}, and a surface defect density of <300 cm^{-2}. The LWIR HgCdTe FPA was fabricated with an n-on-p mesa architecture of grown p–n heterojunctions. The LWIR HgCdTe FPA grown on GaAs also showed a similar detectivity D^* around 6×10^{10} cm $Hz^{1/2}$ W^{-1} and a pixel operability above 99% at 80 K.

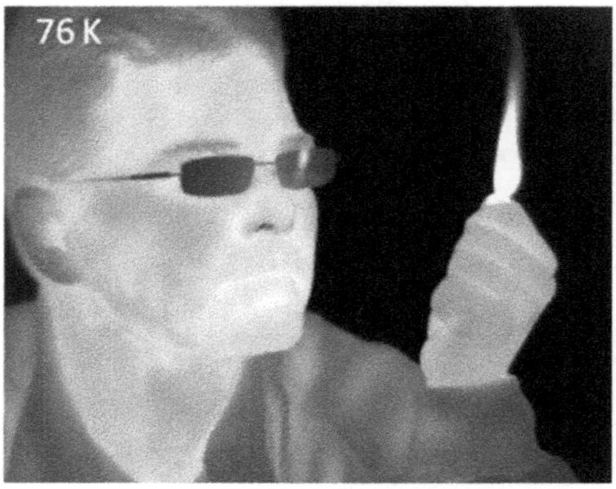

Figure 5.10. Imaging result using the AIM's 1280 × 1040 LWIR HgCdTe FPA grown on GaAs. Reproduced from [32] with permission from Springer Nature.

Researchers from AIM Infrarot-Moduleg GmbH, Germany, also studied LWIR HgCdTe detectors based on HgCdTe grown on GaAs. Wenisch *et al* [31] and Ziegler *et al* [32] fabricated both 640 × 512 and 1280 × 1040 LWIR HgCdTe FPAs (15 μm pitch) based on the HgCdTe on GaAs by using standard planar n-on-p technique at AIM. Under the test conditions F#/2, a 0.15 ms integration time, $T_{OP} = 76$ K, the 1280 × 1040 LWIR HgCdTe FPA showed a mean NEDT of 30.3 mK for 8.8 μm cut-off, and a pixel operability of 99.25%. Figure 5.10 shows the IR image taken by this 1280 × 1040 LWIR HgCdTe FPA on GaAs [32].

In short summary, the device performance of LWIR HgCdTe detectors grown on Si, Ge, and GaAs alternative substrates are much lower than that grown on lattice-matched CdZnTe substrates, and thus cannot meet the requirements for practical industry applications. This is mainly because the threading dislocation density in HgCdTe layers grown on these alternative substrates are typically in the range from mid-10^6 cm^{-2} to low-10^7 cm^{-2}, which is much higher than that (low-10^4 cm^{-2} to low-10^5 cm^{-2}) grown on lattice-matched CdZnTe substrates. There is not much point in studying LWIR detector applications of HgCdTe layers grown on alternative substrates before their EDP is reduced below the critical level of 5×10^5 cm^{-2} as discussed in chapter 3. Therefore, significant effort is needed to achieve this goal in the near future by studying new alternative substrates like GaSb and applying new dislocation reduction methods as discussed in chapter 3.

5.4 Summary

As discussed in previous sections, significant advancement has been made in the IR detector applications of HgCdTe materials grown on alternative substrates such as Si, Ge, and GaAs. SWIR and MWIR HgCdTe detectors and FPAs grown on Si, Ge, and GaAs substrates present a device performance comparable to that grown on CdZnTe substrates. It is noted that although CdTe layers grown on GaSb presented a very low EPD ($\sim 1.5 \times 10^5$ cm^{-2}), no detector or FPA applications have yet been reported (for the reasons previously discussed). The high-performance SWIR and

MWIR HgCdTe detectors grown on alternative substrates have led to practical industry products like the Jupiter product (MWIR HgCdTe FPAs grown on Ge substrates) from Sofradir, France. However, the device performance of LWIR HgCdTe detectors grown on Si, Ge, and GaAs alternative substrates are much lower than that grown on lattice-matched CdZnTe substrates, and thus cannot meet the requirements for practical industry applications. This is mainly due to the fact that the threading dislocation density in HgCdTe layers grown on these alternative substrates are typically in the range from mid-10^6 cm^{-2} to low-10^7 cm^{-2}, which is much higher than that (low-10^4 cm^{-2} to low-10^5 cm^{-2}) grown on lattice-matched CdZnTe substrates. As discussed earlier in this chapter, the EPD in HgCdTe must be controlled below the critical level of 5×10^5 cm^{-2} in order to make high-performance LWIR HgCdTe detectors. Therefore, significant further effort is needed to achieve this goal by studying new alternative substrates and applying new dislocation reduction methods to further drive down the dislocation density in the HgCdTe materials to be below the level of 5×10^5 cm^{-2}, ideally comparable to that of HgCdTe grown on CdZnTe substrates.

References

[1] Love P J, Ando K J, Bornfreund R E, Corrales E, Mills R E, Cripe J R, Lum N A, Rosbeck J P and Smith M S 2002 Large-format infrared arrays for future space and ground-based astronomy applications *Proc. SPIE* **4486** 373–84

[2] Varesi J B, Buell A A, Bornfreund R E, Radford W A, Peterson J M, Maranowski K D, Johnson S M and King D F 2002 Developments in the fabrication and performance of high-quality HgCdTe detectors grown on 4-in. Si substrates *J. Electron. Mater.* **31** 815–21

[3] Reddy M, Peterson J M, Lofgreen D D, Franklin J A, Vang T, Smith E P G, Wehner J G A, Kasai I, Bangs J W and Johnson S M 2008 MBE growth of HgCdTe on large-area Si and CdZnTe wafers for SWIR, MWIR and LWIR detection *J. Electron. Mater.* **37** 1274–82

[4] Bommena R *et al* 2014 High performance SWIR HgCdTe FPA development on silicon substrates *Proc. SPIE* **9070** 907009

[5] Park J H, Pepping J, Mukhortova A, Ketharanathan S, Kodama R, Zhao J, Hansel D, Velicu S and Aqariden F 2016 Development of high-performance eSWIR HgCdTe-based focal-plane arrays on silicon substrates *J. Electron. Mater.* **45** 4620–5

[6] Park J H, Hansel D, Mukhortova A, Chang Y, Kodama R, Zhao J, Velicu S and Aqariden F 2016 Extended short wavelength infrared HgCdTe detectors on silicon substrates *Proc. SPIE* **9974** 99740H

[7] Hu X N *et al* 2017 Large format high SNR SWIR HgCdTe/Si FPA with multiple-choice gain for hyperspectral detection *Proc. SPIE* **10213** 102130E

[8] Zhao J, Li Y H, Yang C Z, Tan Y, Wang X X, Han F Z, Ji R B and Kong J C 2015 MBE growth of Ge-based HgCdTe thin films and their optoelectronic property study *Proc. of the 11th National Conf. of Molecular Beam Epitaxy (China) (Chengdu, Chian)* p 59

[9] Kim J S, An S Y and Suh S H 2002 SWIR diodes of HgCdTe on GaAs substrates grown by metal organic vapor phase epitaxy *Proc. SPIE* **4795** 207–12

[10] Tennant W E, Cabelli S and Spariosu K 1999 Prospects of uncooled HgCdTe detector technology *J. Electron. Mater.* **28** 582–8

[11] Ye Z H, Wu J, Hu X N, Wu Y, Liao Q J, Zhang H Y, Wang J X, Ding R J and He L 2005 A preliminary study on MBE grown HgCdTe two-color FPAs *Proc. SPIE* **5640** 66–73

[12] de Lyon T J *et al* 1998 Molecular-beam epitaxial growth of HgCdTe infrared focal-plane arrays on silicon substrates for midwave infrared applications *J. Electron. Mater.* **27** 550–5

[13] Dhar N K, Zanatta J P, Ferret P and Million A 1999 Characteristics of HgCdTe CdTe hetero-epitaxial system and mid-wave diodes on 2 inch silicon *J. Cryst. Growth* **201/202** 975–9

[14] Maranowski K D, Peterson J M, Johnson S M, Varesi J B, Childs A C, Bornfreund R E, Buell A A, Radford W A, de Lyon T J and Jensen J E 2001 MBE growth of HgCdTe on silicon substrates for large format MWIR focal plane arrays *J. Electron. Mater.* **30** 619–22

[15] Varesi J B, Bornfreund R E, Childs A C, Radford W A, Maranowski K D, Peterson J M, Johnson S M, Giegerich L M, de Lyon T J and Jensen J E 2001 Fabrication of high-performance large-format MWIR focal plane arrays from MBE-grown HgCdTe on 4″ silicon substrates *J. Electron. Mater.* **30** 566–73

[16] Vilela M F, Buell A A, Newton M D, Venzor G M, Childs A C, Peterson J M, Franklin J J, Bornfreund R E, Radford W A and Johnson S M 2005 Control and growth of middle wave infrared (MWIR) HgCdTe on Si by molecular beam epitaxy *J. Electron. Mater.* **34** 898–904

[17] Reddy M *et al* 2011 Molecular beam epitaxy growth of HgCdTe on large-area Si and CdZnTe substrates *J. Electron. Mater.* **40** 1706

[18] Bangs J *et al* 2011 Large format high operability SWIR and MWIR focal plane array performance and capabilities *Proc. SPIE* **8012** 801234

[19] Starr B *et al* 2016 RVS large format arrays for astronomy *Proc. SPIE* **9915** 99152X

[20] He L, Fu X, Wei Q, Wang W, Chen L, Wu Y, Hu X, Yang J, Zhang Q and Ding R 2008 MBE HgCdTe on alternative substrates for FPA applications *J. Electron. Mater.* **37** 1189–99

[21] Zanatta J, Ferret P, Theret G, Million A, Wolny M, Chamonal J and Destefanis G 1998 Heteroepitaxy of HgCdTe (211) B on Ge substrates by molecular beam epitaxy for infrared detectors *J. Electron. Mater.* **27** 542–5

[22] Ferret P, Zanatta J P, Hamelin R, Cremer S, Million A, Wolny M and Destefanis G 2000 Status of the MBE technology at Leti LIR for the manufacturing of HgCdTe focal plane arrays *J. Electron. Mater.* **29** 641–7

[23] Tribolet P *et al* 2006 MWIR focal planes arrays made with HgCdTe grown by MBE on germanium substrates *Proc. SPIE* **6206** 62062F

[24] Dvoretsky S A *et al* 2005 MWIR and LWIR detectors based on HgCdTe/CdZnTe/GaAs heterostructures *Proc. SPIE* **5964** 59640A

[25] Wenisch J, Schirmacher W, Wollrab R, Eich D, Hanna S, Breiter R, Lutz H and Figgemeier H 2015 Evaluation of HgCdTe on GaAs grown by molecular beam epitaxy for high-operating-temperature infrared detector applications *J. Electron. Mater.* **44** 3002–6

[26] Lei W, Gu R J, Antoszewski J, Dell J, Neusser G, Sieger M, Mizaikoff B and Faraone L 2015 MBE growth of mid-wave infrared HgCdTe layers on GaSb alternative substrates *J. Electron. Mater.* **44** 3180–7

[27] Carmody M, Pasko J, Edwall D, Piquette E, Kangas M, Freeman S, Arias J, Jacobs R, Mason W and Stoltz A 2008 Status of LWIR HgCdTe-on-silicon FPA technology *J. Electron. Mater.* **37** 1184–8

[28] Destefanis G and Tribolet P 2007 Advanced MCT technologies in France *Proc. SPIE* **6542** 65420D

[29] Zanatta J, Duvaut P, Ferret P, Million A, Destefanis G, Rambaud P and Vannuffel C 1997 Growth of HgCdTe and CdTe (331) B on germanium substrate by molecular beam epitaxy *Appl. Phys. Lett.* **71** 2984–6

[30] Vilela M, Lofgreen D, Smith E, Newton M, Venzor G, Peterson J, Franklin J, Reddy M, Thai Y and Patten E 2008 LWIR HgCdTe detectors grown on Ge substrates *J. Electron. Mater.* **37** 1465–70

[31] Wenisch J, Eich D, Lutz H, Schallenberg T, Wollrab R and Ziegler J 2012 MBE growth of MCT on GaAs substrates at AIM *J. Electron. Mater.* **41** 2828–32

[32] Ziegler J, Wenisch J, Breiter R, Eich D, Figgemeier H, Fries P, Lutz H and Wollrab R 2014 Improvements of MCT MBE growth on GaAs *J. Electron. Mater.* **43** 2935–40

Chapter 6

Heteroepitaxy of HgCdSe on GaSb—an alternative pathway towards infrared detectors with features of lower cost and larger array format

As discussed in the previous chapters, although significant progress have been made in growing CdTe and HgCdTe layers on alternative substrates such as Si, Ge, GaAs, and GaSb, material quality especially threading dislocation density of the HgCdTe epilayers grown is still not comparable to that grown on lattice-matched CdZnTe substrates due to the large lattice constant and CTE mismatch between CdTe/HgCdTe layer and alternative substrate [1]. Consequently, the applications of HgCdTe grown on alternative substrates are mainly limited to SWIR and MWIR HgCdTe detectors due to the higher EPD numbers (mid-10^6 cm^{-2} to low-10^7 cm^{-2}). In this chapter, we will focus on heteroepitaxy of HgCdSe on GaSb substrate—an alternative pathway towards high performance IR detectors with the needed features of lower cost and larger array format.

6.1 HgCdSe—an emerging infrared material to replace HgCdTe

Recently, ternary II–VI semiconductor alloy HgCdSe grown on its lattice-matched GaSb substrate has emerged as a promising candidate [2] to develop lower cost, larger format and high-performance IR detectors. HgCdSe IR material offers a number of advantages relevant to IR detector applications:

(1) **Wide bandgap tunability:** Similar to $Hg_{1-x}Cd_xTe$, the bandgap of $Hg_{1-x}Cd_xSe$ can be continuously adjusted from 0 to ~1.7 eV by varying the Cd mole fraction [3], thereby enabling wavelength-specific detection across a wide range of the IR spectrum.

doi:10.1088/978-0-7503-3443-3ch6 © IOP Publishing Ltd 2025. All rights, including for text and data mining (TDM), artificial intelligence (AI) training, and similar technologies, are reserved.

(2) **Favorable IR physical properties:** HgCdSe exhibits excellent optoelectronic characteristics such as high electron mobility and potentially long minority carrier lifetime which are comparable to those of HgCdTe [4]. These properties are essential for achieving high-performance IR detectors.
(3) **Compatibility with GaSb substrate:** HgCdSe is nearly lattice-matched to GaSb (0.4% mismatch) according to the lattice constant data (shown in figure 6.1(a)) [5]. Note that GaSb substrate is a mature and commercially available III–V compound semiconductor substrate, and has the features of larger wafer size, lower unit price and higher crystal quality (lower defect density) [3] in comparison to CdZnTe substrates used for growing HgCdTe materials, which are summarized in table 6.1.
(4) **Valence band discontinuity:** Figure 6.1(b) shows the natural valence band offsets of some II–VI semiconductors [6]. It is noted that ZnTe presents nearly zero valence band discontinuity with MWIR and LWIR HgCdSe, which will provide the ideal band configuration for achieving nBn device architectures [7]. In addition, ZnTe has a lattice constant nearly matched with that of HgCdSe. All these make it possible to grow high-quality

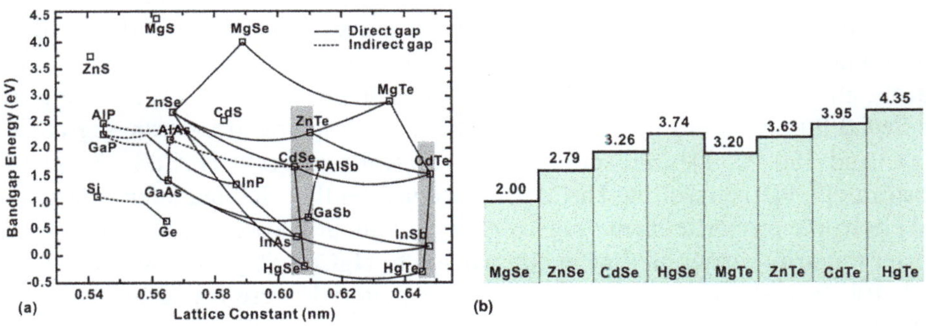

Figure 6.1. (a) Energy bandgap vs lattice constant, and (b) natural valence band offsets (unit: eV) of some semiconductors. The two shaded regions in (a) highlight the HgCdSe and HgCdTe material systems discussed in this book. Panel (a) was reproduced from [5] with the permission from Springer Nature. Panel (b) was reproduced from [6] with the permission of AIP publishing.

Table 6.1. Comparison between GaSb and CdZnTe substrates.

Substrate	Maximum wafer size	Unit price	EPD
GaSb	6 inches in diameter	~\$28 cm^{-2}	~500 cm^{-2}
CnZnTe	8 cm × 8 cm	~\$220 cm^{-2}	Low-10^4 to low-10^5 cm^{-2}

HgCdSe–ZnTe–HgCdSe nBn barrier detector structures and fabricate high-performance IR detectors with lower dark current density and thus higher operating temperature [6, 7].

In addition, HgCdSe belongs to a material family with lattice constants near 6.1 Å (as shown by the shaded area in figure 6.1(a)) including ZnTe, GaSb, and InAs, which will enable the possibility of monolithic integration of multiple detector materials sensitive to different spectral bands on a single chip. This feature is of particular interest for multicolor or multiband IR detection systems. Therefore, HgCdSe/GaSb material system provides a potential alternative material system for developing high-performance IR detectors and their FPAs with features of lower cost and larger array format size.

6.2 MBE heteroepitaxial growth of HgCdSe on GaSb substrates

The first successful MBE growth of HgCdSe epilayers was reported by Lansari *et al* in the 1990s [8]. In their study, ZnTe/CdZnTe was used as the substrate which suffers the same limitations as CnZnTe substrate. Thus, there is no further report on MBE growth of HgCdSe in the subsequent two decades due to the lack of proper epitaxial substrates. With the maturation of GaSb substrate technology, HgCdSe IR materials were re-proposed by the U.S. Army Research Laboratory (ARL) in the early 2010s. Brill *et al* from ARL reported the successful MBE growth of HgCdSe films on lattice-matched GaSb and ZnTe/Si substrates [5]. Their work demonstrated that the alloy composition (x in $Hg_{1-x}Cd_xSe$) could be tuned effectively by adjusting the Se/Cd flux ratio during deposition, with growth temperatures ranging between 80 °C and 130 °C. Despite some continued efforts to grow HgCdSe on ZnTe/Si substrates [2, 9], the resulting HgCdSe films consistently exhibited inferior structural and electronic properties including lower crystal quality, lower carrier mobility, and shorter minority carrier lifetime in comparison to HgCdTe layers grown on CdZnTe [2, 9, 10]. In addition, the as-grown HgCdSe materials were found to have a high background electron concentration (typically $>10^{17}$ cm^{-3} at 77 K) [2]. These results raised doubts about the potential device applications of HgCdSe materials, since carrier mobility, minority carrier lifetime, and background doping level are critical material parameters that determine detector performance. Subsequent studies by Chai *et al* and Zhao *et al* reported the formation of a mixed-phase interfacial layer between the HgCdSe layer and the GaSb substrate, which adversely impacted the crystalline quality of the overgrown HgCdSe epilayer [11, 12].

Based on the previous work from ARL, our UWA group has dedicated significant effort to studying how to achieve high-quality HgCdSe IR materials on GaSb substrates with physical properties that are suitable for making high-performance IR detectors. Generally, our UWA group introduced a ZnTe buffer layer between the GaSb substrate and the HgCdSe epilayer to suppress the atom interdiffusion, avoid the formation of mixed phases around the interface areas, and isolate the HgCdTe layer from the conductive GaSb substrate [4, 13].

The reported MBE growth process of HgCdSe layers on GaSb substrate is as follows [4]:

The HgCdSe epilayers were grown on epi-ready GaSb (211) B substrates in a Riber 32P MBE system. Hg (7N), Cd (6N), Se (5N5), Zn (6N), and Te (6N) were used as the source materials, and standard effusion cells were used to evaporate the source materials without cracking. Following thermal desorption of the native oxide from the GaSb substrate surface at 580 °C, the substrate temperature was reduced to 320 °C for growing a thin ZnTe buffer layer (200–300 nm thick). After that, the substrate temperature was reduced to the required temperature (70 °C–120 °C) for growing the HgCdSe epitaxial layers. After the growth of HgCdSe, the samples were cooled down to room temperature without any background Hg flux. For the thermal desorption of native oxide from GaSb substrate surface, GaSb substrates were held at 580 °C for 5 min without the protection of background Sb flux. This is because Sb is an effective p-type dopant for HgCdSe, and thus can cause serious cross-contamination to the II–VI MBE growth chamber if introducing background Sb flux during the thermal desorption of native oxide. During the growth of HgCdSe, a large Hg BEP (~low-10^{-4} Torr) was used due to the very low sticking coefficient of Hg, while much lower Se BEP (~low-10^{-6} Torr) and Cd BEP (~mid-10^{-7} Torr) were used, which are comparable to those used for the MBE growth of HgCdTe [14]. To study the growth mechanism and physical properties of HgCdSe materials, the epilayers were grown with different compositions as well as at different substrate temperatures (70 °C–120 °C). The HgCdSe composition was tuned by varying Se/Cd BEP ratio, which is similar to that reported previously by other researchers [2, 5, 15].

Figure 6.2(a) shows a typical RHEED pattern obtained during the growth of HgCdSe layers, demonstrating uniform, long and narrow streaks indicative of a high crystal quality as well as a smooth growth front [14]. Figures 6.2(b)–(d) present cross-sectional SEM images of a HgCdSe sample as well as the EDX spectra collected from the white dashed rectangle areas marked in the SEM image. As indicated by their EDX spectra, the light and dark regions in figure 6.2(b) represent the HgCdSe layer and GaSb substrate, respectively, with the thickness of the HgCdSe layer determined to be 14.1 μm. Note that in comparison with the HgCdSe layer and GaSb substrate the ZnTe buffer layer (300 nm) is very thin, and thus appears as the highly contrasted bright thin layer in figure 6.2(b). A sharp interface is observed between the HgCdSe layer and the GaSb substrate, indicating a high-quality MBE growth without any atomic interdiffusion from the GaSb substrate into the HgCdSe layer. This sharp interface without atomic interdiffusion can be attributed to the low growth temperature and the diffusion-blocking effect of the ZnTe buffer layer, which also was observed in the TEM study of ZnTe layers grown on GaSb substrates reported [16]. Figure 6.2(e) presents a typical XRD rocking curve of a HgCdSe epilayer, with an XRD FWHM of 116 arcsec, which indicates a high-quality MBE growth, given that the starting GaSb substrate had a FWHM of 37 arcsec as shown in the inset of figure 6.2(e). Another important material parameter for IR materials is the dislocation density, which has a significant impact

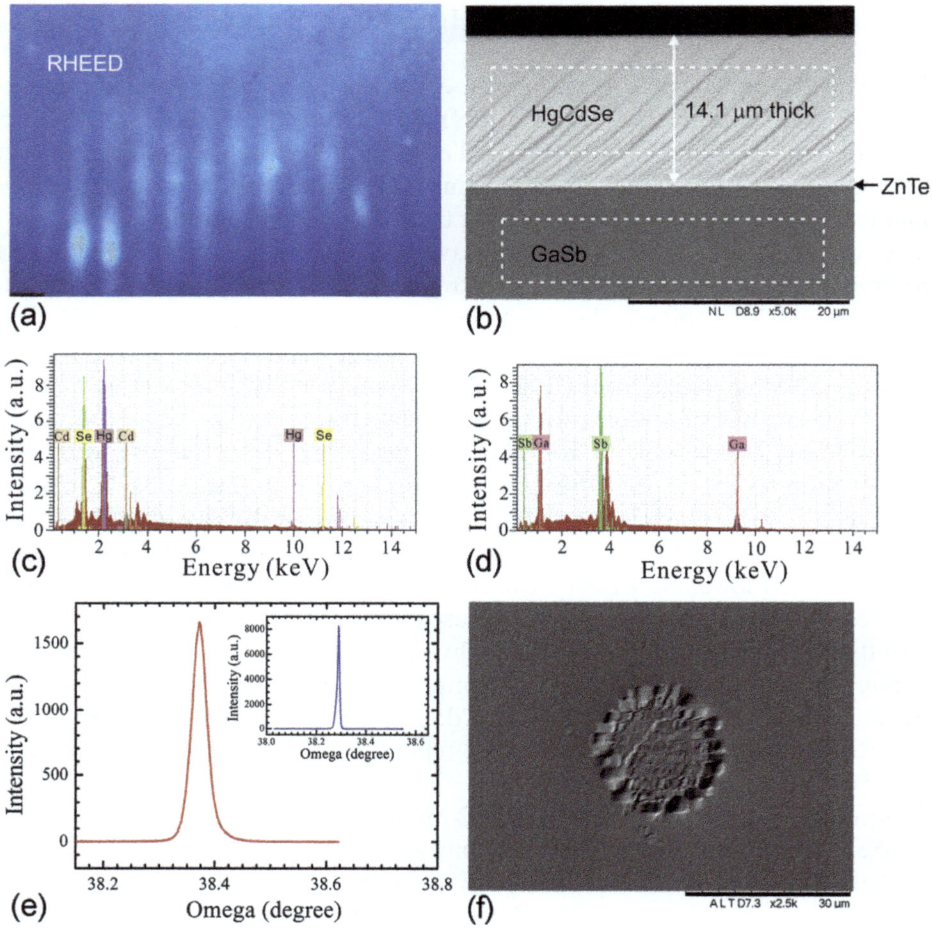

Figure 6.2. (a) Representative RHEED pattern during HgCdSe growth; (b) cross-sectional SEM image; (c) EDX spectrum collected from the areas identified by the white dashed rectangle in the upper part of the cross-sectional SEM image; (d) EDX spectrum collected from the areas identified by the white dashed rectangle in the lower part of the cross-sectional SEM image; (e) XRD rocking curve of a HgCdSe sample; and (f) SEM image of large defect structure on HgCdSe sample surface after etch pit density etching. The inset of (e) shows the XRD rocking curve of the GaSb substrate. (Reprinted from [4], Copyright © 2018, with permission from Elsevier B.V. All rights reserved.)

on the yield and performance of subsequently fabricated detector devices. To study the dislocation density in HgCdSe materials, EPD measurements were undertaken using a 30 s etch for revealing dislocations in CdSe substrates [17]. As shown in figure 6.2(f), circular pits of 10–20 μm in size with a density of $\sim 2.2 \times 10^3$ cm^{-2} were observed on the surface of the HgCdSe epitaxial layers. This EPD number is three orders of magnitude lower than that of HgCdSe materials grown on ZnTe/Si substrates ($\geq$mid-10^6 cm^{-2}) reported previously [18], and comparable to that of the

GaSb substrate (~low-10^3 cm^{-2}). Such a low EPD is mainly due to the nearly lattice-matched growth of HgCdSe and ZnTe on GaSb, in comparison to the lattice-mismatched growth of HgCdSe and ZnTe on Si reported previously [18].

Figure 6.3(a) shows the 80 K fourier transform infrared (FTIR) transmission spectra of several HgCdSe samples grown with different Se/Cd BEP ratios. By varying the Se/Cd ratio from 6 to 7.8, the cut-off wavelength of the HgCdSe material can be tuned from 3.9 to 10.4 μm due to the corresponding change of HgCdSe alloy composition. The x value of the $Hg_{1-x}Cd_xSe$ alloy was determined to be 0.37, 0.21, 0.19 and 0.18 for HgCdSe materials with a cut-off wavelength of 3.9, 6.6, 9 and 10.4 μm at 80 K, respectively. This tuneable-wavelength capability is very similar to that of HgCdTe, where the cut-off wavelength/composition can also be tuned by varying the Te/CdTe BEP ratio [19]. In order to gain a better understanding of the bandgap behavior of the HgCdSe alloy, temperature-dependent FTIR transmission spectroscopy was undertaken on HgCdSe materials with x values of 0.37, 0.21 and 0.19. Figure 6.3(b) presents the HgCdSe cut-off wavelength measured at different temperatures. It is observed that the cut-off wavelength shows a consistent blue shift with increasing temperature, which is very similar to the behavior of HgCdTe [20], and is caused by electron-phonon interaction (mainly acoustic phonons) in the material [21]. For the HgCdSe samples studied in figure 6.3(b), the observed blue shift of cut-off wavelength with increasing temperature is much greater for low x-value alloys, that is, for alloys with a higher HgSe/CdSe ratio [22]. Based on the previously reported empirical equation relating cut-off wavelength to alloy composition and temperature [23], the cut-off wavelengths of the three HgCdSe samples ($x = 0.37$, 0.21 and 0.19) at different temperatures were also calculated, the results of which are plotted in figure 6.3(b). As indicated in figure 6.3(b), the change of the cut-off wavelength of HgCdSe with increasing temperature measured in this work does not follow the reported empirical equation in reference [23]. An example is that the

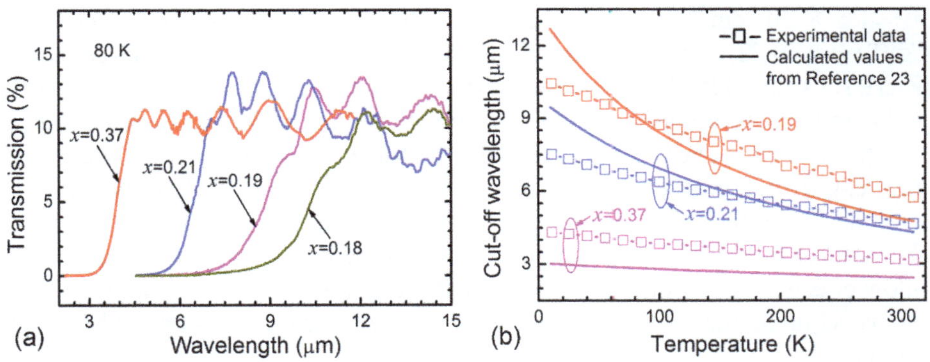

Figure 6.3. (a) 80 K FTIR transmission spectra of HgCdSe materials with different composition; (b) temperature dependent cut-off wavelength of HgCdSe materials with x vales of 0.37, 0.21 and 0.19. (Reprinted from [4], Copyright © 2018, with permission from Elsevier B.V. All rights reserved.)

observed rate at which the blue-shift of cut-off wavelength varies with increasing temperature is significantly different from those calculated based on the empirical equation relating cut-off wavelength to alloy composition and temperature [23], especially for the low x-value alloys. Considering that the HgCdSe materials reported in reference [23] were grown via the Bridgman method; this suggests that the material growth method has a significant influence on the electronic bandgap behavior of HgCdSe materials, the main reason for which could be attributed to the high background electron concentration in the HgCdSe materials grown and will be investigated in detail later in this chapter.

Apart from the structural properties and electronic bandgap, the two critical physical parameters of semiconductor materials which directly impact on the quantum efficiency of IR detectors are the carrier mobility and minority carrier lifetime. To determine the carrier mobility, mobility spectrum analysis (MSA) with Hall measurements were undertaken on the HgCdSe samples, which allows the conductivity contribution of the HgCdSe epitaxial layer to be separated from that of the conductive GaSb substrate [24]. Figures 6.4(a) and (b) show the representative MSA spectra of a GaSb substrate and an as-grown HgCdSe sample ($x = 0.18$, 10.4 μm cutoff wavelength at 80 K) measured at 132 K, respectively. With reference to the HgCdSe epilayer, the dominant carrier species is a high-mobility electron with concentration of 1.9×10^{16} cm^{-3}, and a mobility of approximately 6.8×10^4 cm^2 V^{-1} s^{-1}. To study the temperature behavior of electron mobility and background electron concentration, temperature-dependent MSA measurements were also carried out on this HgCdSe sample ($x = 0.18$). Figures 6.4(c) and (d) show the electron mobility and background electron concentration measured at different temperatures. It is observed that with reducing temperature the electron mobility increases significantly while the background electron concentration decreases significantly, especially at low temperatures. At 80 K, the electron mobility of the dominant carrier is measured to be 1.3×10^5 cm^2 V^{-1} s^{-1} and the background electron concentration to be 1.6×10^{16} cm^{-3}. This electron mobility in our as-grown HgCdSe material is significantly higher than that reported for corresponding HgCdSe in previous work, which was in the range of 2.5×10^4 to 3×10^4 cm^2 V^{-1} s^{-1} [2], and is comparable to that ($\sim 1.5 \times 10^5$ cm^2 V^{-1} s^{-1}) of corresponding long-wave infrared (LWIR) $Hg_{1-x}Cd_xTe$ ($x = 0.22$, ~10.7 μm cut-off wavelength at 80 K) grown on lattice-matched CdZnTe substrates [25, 26]. The results clearly confirm the high crystal quality of the HgCdSe epitaxial layers grown in this work, and the n-type behavior can be attributed to the existence of native Se vacancies and other impurities [2]. The electron concentration in the as-grown HgCdSe epitaxial layers presented in this work is one order of magnitude lower than that reported in previous studies [2]. Note that ideally, the background carrier concentration needs to be controlled below the low-10^{15} cm^{-3} range to allow effective and controllable extrinsic doping in the 10^{16} cm^{-3} range or above, which is required for fabricating p–n junctions for high-performance photovoltaic detectors.

To determine the minority carrier lifetime in our HgCdSe epitaxial layers, photoconductive decay measurements were undertaken. Figure 6.4(e) presents the photoconductive decay curve of an as-grown HgCdSe sample ($x = 0.18$) measured at 80 K. The minority carrier lifetime is determined to be 2.22 μs, by fitting the

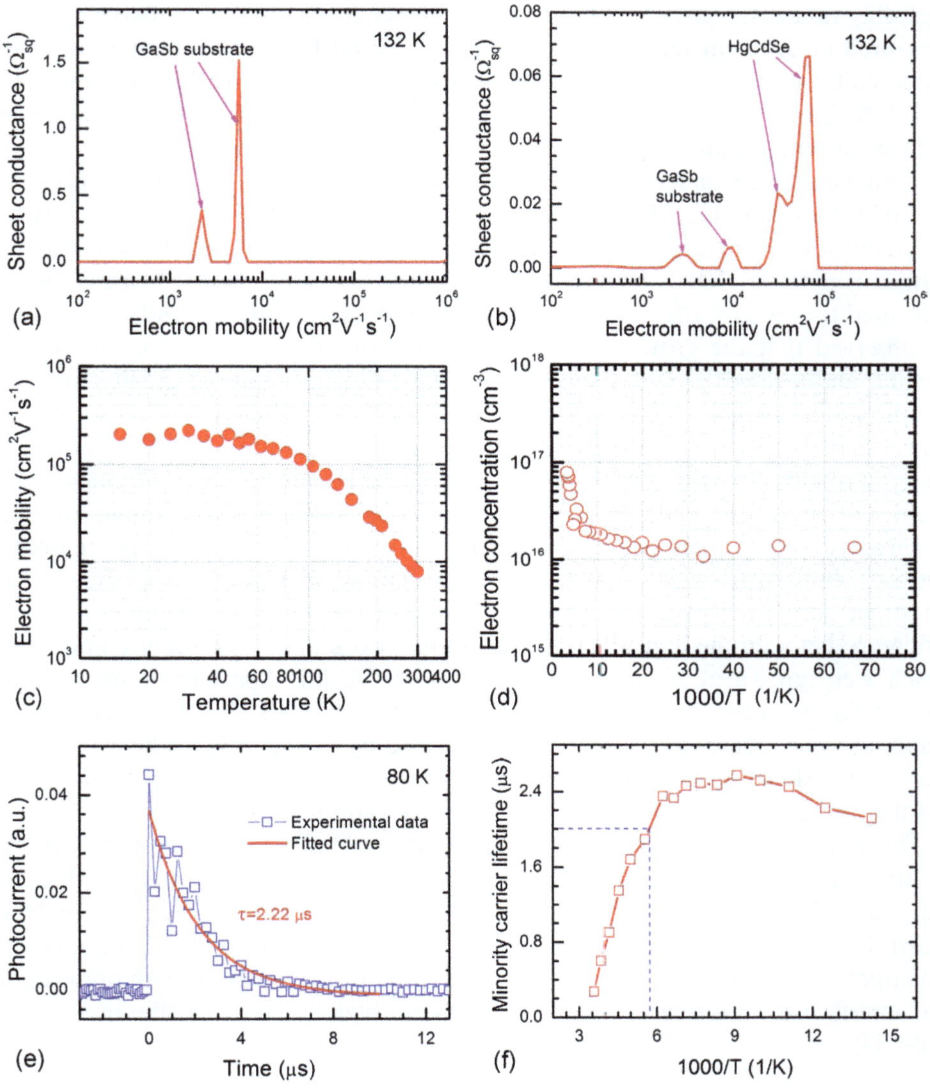

Figure 6.4. Experimental results from HgCdSe ($x = 0.18$) grown at 70 °C on GaSb substrate: electron mobility spectrum at 132 K of (a) GaSb substrate alone and (b) HgCdSe-on-GaSb sample; temperature dependence of HgCdSe (c) electron mobility and (d) electron concentration; results of (e) 80 K photoconductive decay measurement, and (f) temperature dependent minority carrier lifetime. (Reprinted from [4], Copyright © 2018, with permission from Elsevier B.V. All rights reserved.)

experimental data with an exponential of the form $y = y_0 + A_1 \times \exp(-x/\tau)$, as reported in previous work [27]. This 2.22 μs lifetime is significantly longer than that (∼132 ns) reported for corresponding HgCdSe in previous work [28], and is comparable to that (∼2 μs) of corresponding LWIR $Hg_{1-x}Cd_xTe$ ($x = 0.22$, ∼10.7 μm cut-off wavelength at 80 K) materials grown on lattice-matched CdZnTe substrates [26]. Once again, this confirms the high crystal quality of the

HgCdSe materials grown. Figure 6.4(f) presents the minority carrier lifetime measured at different temperatures ranging from 80 K to room temperature. It is observed that a >2 μs minority carrier lifetime can be maintained approximately up to 175 K, indicating the potential for fabricating IR detectors that can operate with reduced cooling requirements.

To have a better understanding of the growth mechanism and material quality of HgCdSe materials, HgCdSe samples with similar x values (~0.18) were grown at different growth temperatures ranging from 70 °C to 120 °C [4, 13]. Table 6.2 lists the surface roughness, XRD FWHM, electron mobility, and minority carrier lifetime measured for the HgCdSe materials grown at 70 °C, 80 °C, 100 °C and 120 °C. It can be observed that the growth temperature has a significant impact on the material quality, and a lower growth temperature leads to higher material quality: smoother sample surface, smaller XRD FWHM, higher electron mobility, and longer minority carrier lifetime. For our MBE facility, the growth temperature of 70 °C gives the best material quality for HgCdSe. This may be because Hg atoms have very low sticking coefficient at higher temperatures [14], and thus lower growth temperatures are required to ensure there are sufficient Hg atoms on the material growth front to nucleate and form single crystal material, resulting in higher crystal quality and smoother sample surface. This is also the main reason for the material quality difference between the HgCdSe materials in this work and those reported previously which were grown at higher temperature ~100 °C [2]. Apart from the optimum growth temperature, the nearly lattice-matched growth of ZnTe and HgCdSe on GaSb also contributes to the higher material quality of HgCdSe in this work due to the lower dislocation density formed in the materials, in comparison to the lattice-mismatched growth of ZnTe and HgCdSe on Si reported previously [18].

The initial study at UWA led to high-quality HgCdSe IR materials with key electrical properties that were comparable to those of the counterpart HgCdTe materials. However, the overall material quality and electrical properties of MBE-grown HgCdSe epilayers are still inferior to those of HgCdTe on CdZnTe, especially the larger XRD FWHM (~116 arcsec). Therefore, further MBE growth optimization and interface engineering are needed to achieve high-quality HgCdSe materials on GaSb [3].

Table 6.2. XRD FWHM, electron mobility, and minority carrier lifetime of HgCdSe (x = 0.18) materials grown at different growth temperatures.

Growth temperature (°C)	70	80	100	120
RMS surface roughness (nm)	2.7	3	3.95	5.43
XRD FWHM (arcsec)	116	151	216	234
Electron mobility at 132 K (× 10^4 cm^2 $V^{-1}s^{-1}$)	6.8	2.0	2.2	1.5
Minority carrier lifetime at 80 K (μs)	2.2	1.35	1.1	0.71

6.2.1 Thermal desorption optimization of GaSb substrate surface

As stated previously in the MBE growth process for growing HgCdSe layers on GaSb, no Sb background flux was applied during the thermal desorption of GaSb substrates due to the described limitations. As a result, the GaSb substrates might be not properly desorbed, leaving a rough and even incompletely desorbed sample surface for the subsequent growth of HgCdSe layers. This would degrade the crystalline quality and physical properties of the HgCdSe layers such as the large XRD FWHM observed for the HgCdSe samples grown. Therefore, over the past several years, our UWA group has devoted significant effort to optimizing thermal desorption of GaSb substrates [14, 29]. It is known that an ideal substrate surface should be atomically smooth, mirror-like, and free from any native oxides or contamination prior to growth [30]. It has been reported that complete desorption of native oxides from GaSb surfaces can be achieved at substrate temperatures between 500 °C and 550 °C [31–33]. However, Schwartz *et al* [34] observed that excessively high thermal budgets may trigger undesirable surface reactions such as:

$$Sb_2O_3 + 2GaSb \rightarrow Ga_2O_3 + 4Sb \tag{6.1}$$

This reaction is known to cause surface degradation and roughening, potentially leading to morphological undulations in the grown films that significantly degrade their structural quality [35].

To establish an optimal thermal desorption protocol for GaSb substrates prior to HgCdSe epitaxial growth, a systematic investigation was carried out in this study. GaSb substrates were thermally desorbed/cleaned at different temperatures—500 °C, 510 °C, 520 °C, 530 °C, and 540 °C, while keeping all other growth conditions constant [36]. Figure 6.5 presents the *in-situ* RHEED patterns obtained from the GaSb surfaces after 15 min thermal desorption at each temperature. The RHEED pattern in figure 6.5(a), recorded after desorption at 500 °C, shows a spotty morphology, indicating incomplete oxide removal. In contrast, the RHEED pattern in figure 6.5(b), corresponding to 510 °C, exhibits a long, well-defined streaky characteristic suggesting a clean and atomically flat surface. However, the further increase of the cleaning temperature to 520 °C, 530 °C, and 540 °C results in the bold spotty patterns, as seen in figures 6.5(c)–(e). These bold patterns are attributed to surface roughening possibly induced by the reaction described in equation (6.1). From this study, it is observed that thermal cleaning at 510 °C for 15 min provides the optimal condition for achieving a clean, oxide-free, and smooth GaSb substrate surface with the Riber 32P MBE system at UWA.

To further validate the results of *in-situ* RHEED analysis, XRD measurements were also undertaken on the HgCdSe epilayers grown on GaSb substrates thermally desorbed at different temperatures for 15 min. Figure 6.6 presents the XRD rocking curves for these samples [36]. It is observed that the HgCdSe layer grown on the GaSb substrate thermally desorbed at 510 °C shows the smallest XRD FWHM, approximately 65 arcsec, indicative of the highest crystalline quality. In contrast, the HgCdSe sample prepared with a thermal desorption temperature of 500 °C shows a significantly larger XRD FWHM of 492 arcsec, and the XRD peak notably shifts

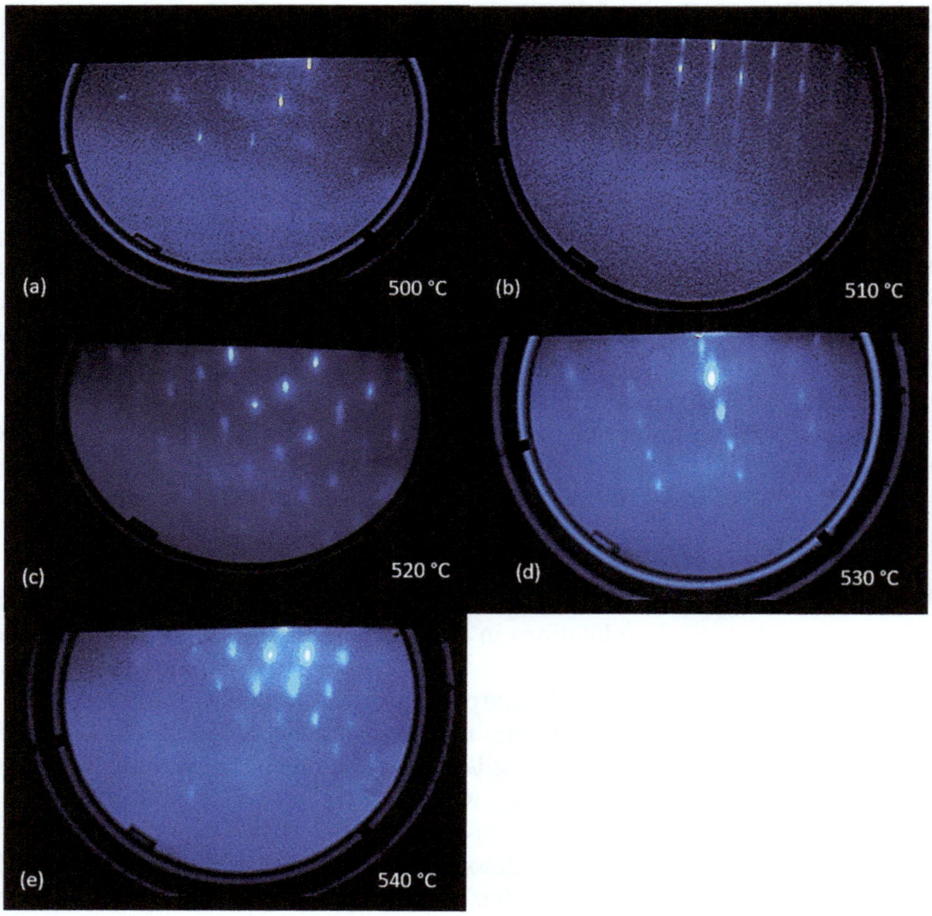

Figure 6.5. Representative RHEED patterns of the GaSb substrate surfaces after being thermally cleaned for 15 min at (a) 500 °C, (b) 510 °C, (c) 520 °C, (d) 530 °C, and (e) 540 °C. (Reproduced from [36]. CC BY 4.0.)

relative to the other samples. This suggests that the HgCdSe film is either highly disordered or partially amorphous, likely due to incomplete removal of surface oxides from the GaSb substrate prior to HgCdSe growth. At elevated thermal cleaning temperatures of 520 °C, 530 °C, and 540 °C, the XRD FWHM value of the corresponding HgCdSe layers also increases. This degradation in crystal quality is consistent with the earlier hypothesis that excessive thermal energy may trigger surface roughening or initiate chemical reactions, leading to morphological non-uniformities during epitaxy. These surface irregularities translate directly into poorer structural coherence in the overgrown layers, as reflected in the larger XRD FWHM values. Overall, both the RHEED and XRD results show that the thermal desorption at 510 °C for 15 min is the optimum thermal desorption condition for GaSb substrate with the Riber 32 P MBE system at UWA. Note that XRD FWHM of 65 arcsec for HgCdSe suggests high-quality MBE growth considering the 37 arcsec XRD FWHM for GaSb substrates.

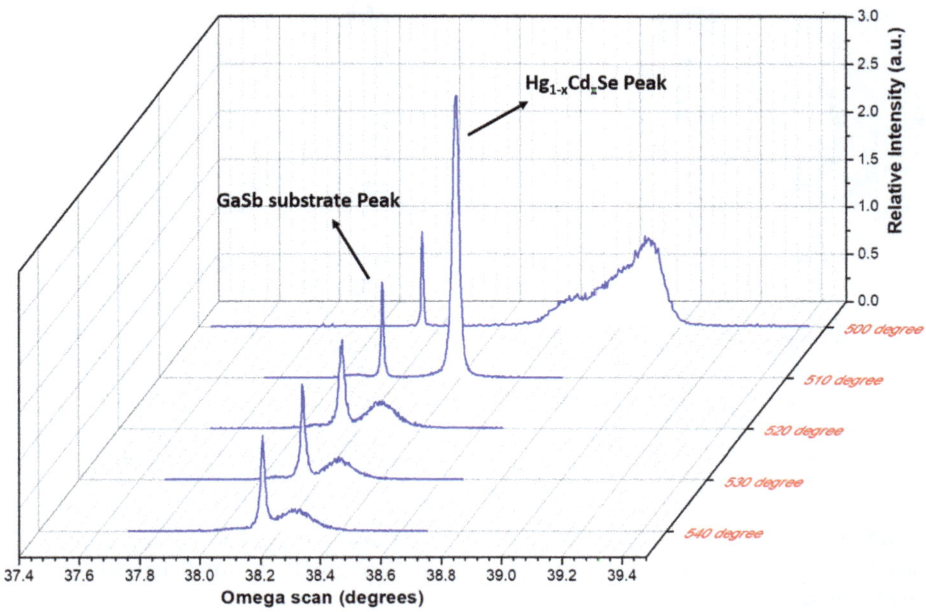

Figure 6.6. XRD rocking curves of HgCdSe/ZnTe/GaSb samples with a thermal desorption at 500 °C, 510 °C, 520 °C, 530 °C, and 540 °C for 15 min. The arrows indicate the typical peaks represented by HgCdSe layer and GaSb substrate, respectively. (Reproduced from [36]. CC BY 4.0.)

6.2.2 ZnTe buffer layer optimization [3]

As discussed before, ZnTe buffer layer was typically used for growing HgCdSe on GaSb substrates. Despite its near lattice matching, ZnTe presents a tiny residual mismatch with GaSb (~0.13%), which can lead to misfit strain accumulation and subsequent dislocation formation in the overlying layers. As such, the thickness of the ZnTe buffer layer must be optimized to maintain the fully strained state with GaSb substrate. To determine the optimal buffer layer thickness, ZnTe films with different thicknesses—150, 300, 400, 600, and 1000 nm, were grown on GaSb substrates without the subsequent deposition of HgCdSe. These samples were analyzed using XRD reciprocal space mapping (RSM) to quantify the degree of lattice relaxation and strain distribution. RSM data were acquired by performing a series of ω–2θ scans at varying ω values. The resulting diffraction data were converted into reciprocal space coordinates (q_x, q_y) using the following transformations [37]:

$$q_x = \frac{\cos\omega - \cos(2\theta - \omega)}{\lambda} \tag{6.2}$$

and

$$q_y = \frac{\sin\omega + \sin(2\theta - \omega)}{\lambda}, \tag{6.3}$$

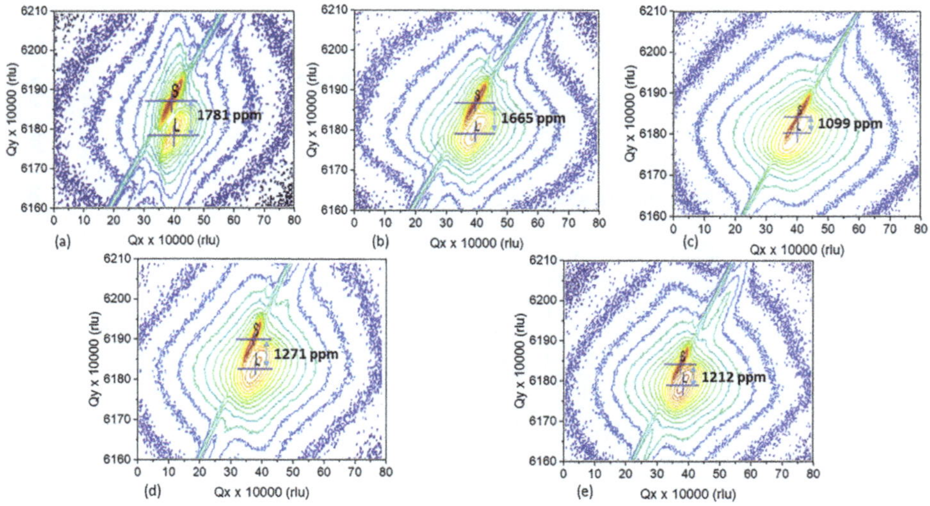

Figure 6.7. XRD RSM data for ZnTe buffer layers grown on GaSb substrates with a thickness of 150 nm (a), 300 nm (b), 400 nm (c), 600 nm (d), and 1000 nm (e). (Reproduced from [3]. © 2019 Chinese Physical Society and IOP Publishing Ltd. All rights reserved.)

where ω and θ are the incident and diffracted angles, respectively, λ is the x-ray wavelength, and q_x, q_y correspond to the lateral and vertical reciprocal lattice vectors, respectively. All RSM measurements were performed along the symmetric (422) reflection, corresponding to the crystallographic growth orientation of the epilayers. Figure 6.7 presents the RSM results for the series of ZnTe buffer layers. These data provide critical insight into the strain relaxation behavior as a function of ZnTe thickness and can be used to determine the optimal buffer thickness for enabling high-quality HgCdSe growth.

In a symmetrical RSM, the vertical separation between diffraction peaks provides direct information on the lattice mismatch and strain status along the growth direction. Given that the lattice constant of ZnTe (6.105 Å) is slightly larger than that of GaSb (6.095 Å), ZnTe epilayers grown on GaSb substrates experience in-plane compressive strain and corresponding tensile strain along the out-of-plane (growth) direction, as illustrated schematically in figure 6.8. During MBE growth, ZnTe is deposited in a layer-by-layer fashion. As growth proceeds, misfit strain resulting from the lattice mismatch between ZnTe and GaSb accumulates. When the film thickness reaches a critical threshold or 'critical thickness,' the strain energy within the layer surpasses the limit that can be elastically sustained by the material system. At this point, the system undergoes strain relaxation via the formation of misfit dislocations, thereby lowering the total elastic energy of the heterostructure. The value of this critical thickness is primarily determined by the magnitude of lattice mismatch: a smaller mismatch yields a larger critical thickness. For high-quality growth of HgCdSe on GaSb via a ZnTe buffer layer, it is therefore essential to maintain the ZnTe layer thickness below its critical thickness to prevent dislocation generation via strain relaxation that would compromise the structural integrity of the subsequent HgCdSe epilayer.

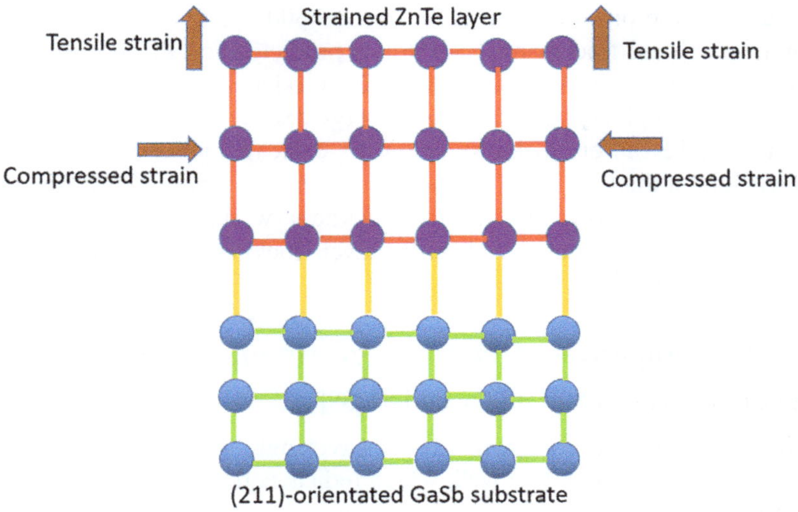

Figure 6.8. Schematic diagram of strained ZnTe layer grown on GaSb substrate. (Reproduced from [3]. © 2019 Chinese Physical Society and IOP Publishing Ltd. All rights reserved.)

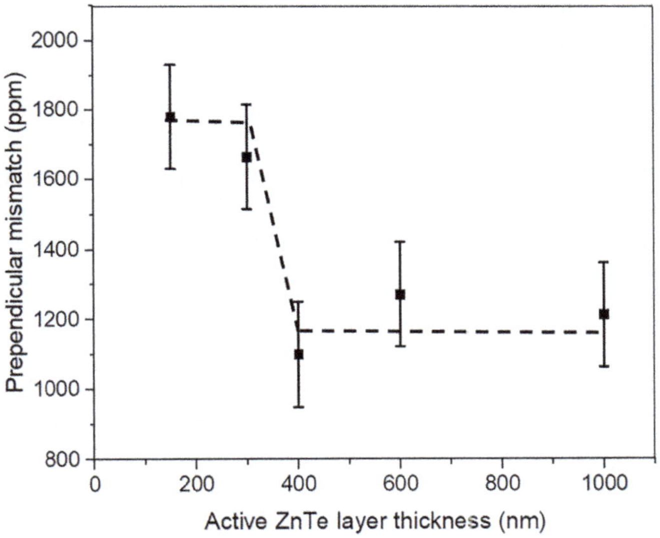

Figure 6.9. Misfit strain along the growth direction (vertically) in ZnTe buffer layer samples with different ZnTe layer thicknesses. (Reproduced from [3]. © 2019 Chinese Physical Society and IOP Publishing Ltd. All rights reserved.)

Figure 6.9 presents the vertical (out-of-plane) misfit strain values derived from the RSM results shown in figure 6.7. A notable change in mismatch is observed between ZnTe buffer layers of 300 and 400 nm thickness, where the measured perpendicular mismatch level decreases from ~1600 to ~1000 ppm. This sudden reduction strongly suggests the onset of relaxation beyond 300 nm thickness, consistent with the theoretical critical thickness of ~316 nm predicted for ZnTe grown on GaSb (211)B substrates using the Matthews–Blakeslee model [38]. To minimize dislocation density and ensure high crystal quality of the HgCdSe active layers, it is important to maintain the ZnTe buffer thickness below its critical thickness. At UWA, ZnTe buffer layers with a thickness of ⩽200 nm were used for growing HgCdSe, providing a practical margin below the critical thickness while still ensuring sufficient operation tolerance. This optimized buffer layer thickness facilitates the growth of high-quality HgCdSe epilayers.

6.3 Material properties of HgCdSe grown on GaSb substrates

6.3.1 Optical bandgap of HgCdSe materials [39]

As shown in figure 6.3, the change of cut-off wavelength with increasing temperature did not follow the empirical data well, which indicates there is a divergence between optical bandgap and intrinsic bandgap. To better understand this, a $Hg_{1-x}Cd_xSe$ ($x = 0.21$) sample was selected for more detailed discussion and analysis. Figure 6.10 presents the temperature dependence of the optical bandgap E_{op}, of this $Hg_{0.79}Cd_{0.21}Se$ sample extracted from its FTIR transmission/absorption spectra [39], which shows an approximately linear increase with increasing temperature. However, at temperatures below 200 K, there is a clear deviation between the measured E_{op} and the intrinsic bandgap E_g. Note E_g values were calculated using the empirical equation developed by Summers and Broerman [40]. The divergence between E_{op} and E_g at low temperatures can be attributed to conduction band degeneracy resulting from the high background doping level. The MBE-grown $Hg_{0.79}Cd_{0.21}Se$ sample in this study exhibited a background electron concentration (n_B) of approximately 3.5×10^{16} cm^{-3} at 77 K [41]. At this doping level, the Fermi level (E_F) is located within the conduction band, leading to the occupation of

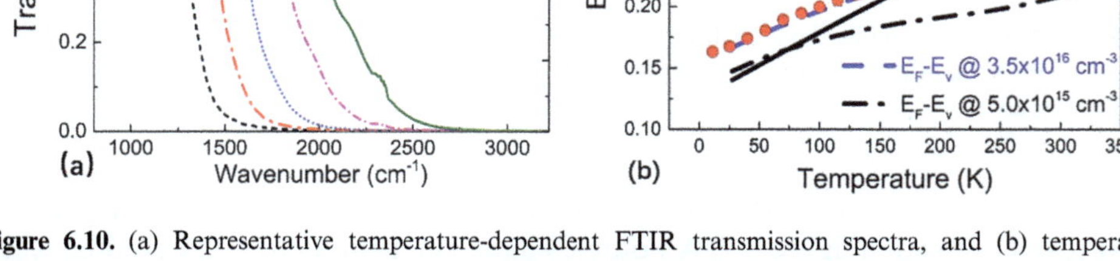

Figure 6.10. (a) Representative temperature-dependent FTIR transmission spectra, and (b) temperature dependence of the measured optical bandgap of the $Hg_{0.79}Cd_{0.21}Se$ sample compared to the empirical calculations based on [40], and the calculated Fermi levels for two different electron concentrations. (Reproduced from [39] with permission from Springer Nature.)

electronic states near the conduction band edge (E_C). These occupied states do not contribute to optical absorption, resulting in an upward shift of the apparent absorption edge. Consequently, the measured E_{op} reflects the energy difference between the valence band maximum and E_F, rather than the fundamental bandgap E_g. This phenomenon is recognized as the Burstein–Moss (BM) effect, commonly observed in degenerated narrow-gap semiconductors such as HgCdTe [42]. Therefore, to enable the fabrication of high-performance IR detectors using HgCdSe, it is essential to suppress the background doping level. As demonstrated in figure 6.10, reducing n_B from 3.5×10^{16} cm^{-3} to 5.0×10^{15} cm^{-3} can significantly mitigate the BM shift, eliminating the gap between E_{op} and E_g at 77 K. This result highlights the critical importance of background doping control being below the level of 5.0×10^{15} cm^{-3}.

6.3.2 Optical absorption coefficient of HgCdSe materials [39]

Optical absorption coefficient is an important physical parameter for IR detectors which determine whether the IR light can be effectively absorbed or not. Chu *et al* [43] proposed an empirical model for the intrinsic absorption coefficient ($E > E_g$, direct band-to-band transition), which provides an effective and practical fitting approach. The model defines the absorption coefficient in the intrinsic (Kane theory) region as:

$$\alpha = \alpha_g \exp\,[\beta(E - E_g)]^{1/2}, \tag{6.4}$$

where α_g is the absorption coefficient at the bandgap E_g, and β is an empirical fitting parameter dependent on both composition (x) and temperature (T).

For the sub-bandgap region ($E < E_g$), the absorption coefficient follows the exponential Urbach tail, described by:

$$\alpha = \alpha_0 \exp\left[\frac{\delta(E - E_0)}{kT}\right], \tag{6.5}$$

where δ is a fitting constant, k is the Boltzmann constant, and α_0 and E_0 are constants independent of energy and temperature. According to Urbach's rule, absorption edges measured at different temperatures should converge at the characteristic point (E_0, α_0). The point of transition between the Urbach tail and intrinsic absorption defines E_g and α_g, with continuity ensured by $\delta/kT = (\ln\alpha_g - \ln\alpha_{g0})/(E_g - E_0)$. The following empirical expressions can be used to fit the absorption spectra of the $Hg_{1-x}Cd_xSe$ samples:

$$\ln \alpha_0 = -21.03 + 52.27x \tag{6.6}$$

$$E_0 = -0.307 + 1.909x \tag{6.7}$$

$$\alpha_g = (-46.14 + 735.94x)\exp(0.0058T) \tag{6.8}$$

$$\beta = 185.82 + 0.227T + (150.9 - 3.205T)x \tag{6.9}$$

As indicated by these relationships, the absorption coefficient α_g slightly increases with both temperature and Cd composition. These models are valid under non-degenerated conditions, where the influence of the Burstein–Moss effect is negligible.

6.3.3 Mechanical properties of HgCdSe materials [44, 45]

Mechanical properties of materials are essential for their manufacturing process. Therefore, it is important to study and understand the mechanical properties of HgCdSe materials as well as their benchmarking with HgCdTe materials. Our UWA group undertook a detailed study on the hardness and elastic modulus of $Hg_{1-x}Cd_xSe$ samples grown on GaSb (211)B substrates ($x = 0, 0.19, 0.24, 0.31, 0.36, 0.46, 1$) by nanoindentation technique [44, 45]. Detailed nanoindentation experiments can be found in references [44, 45]. The average hardness values were determined to be 0.56 ± 0.04, 0.67 ± 0.05, 0.77 ± 0.09, 0.93 ± 0.05, 1.02 ± 0.04, 1.09 ± 0.07, and 1.63 ± 0.14 GPa for $x = 0, 0.19, 0.24, 0.31, 0.36, 0.46$, and 1, respectively. These values are plotted in figure 6.11 as a function of alloy composition [44, 45]. The measured hardness values for HgSe ($x = 0$) and CdSe ($x = 1$) are consistent with previously reported literature values—approximately 0.55 GPa for HgSe and 1.35 GPa for CdSe [46, 47]. As shown in figure 6.11, the hardness exhibits a monotonic increase with increasing Cd composition, which is consistent with trends observed in the prior studies on this material system [48, 49]. The observed increase in hardness with increasing Cd composition is primarily

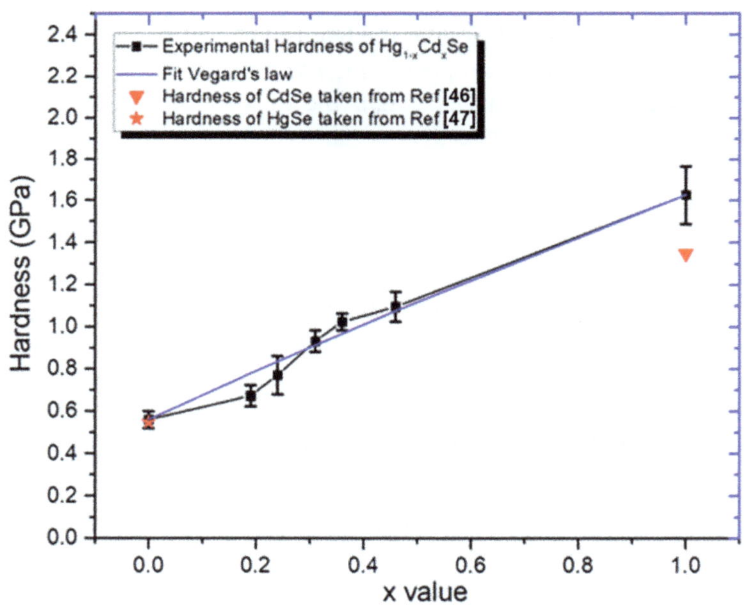

Figure 6.11. The nanoindentation measured H value as a function of x for $Hg_{1-x}Cd_xSe$ thin films grown on GaSb (211) B substrates. *Note*: Black squares refer to the measured average H values, the blue line refers to the H values calculated according to Vegard's law, while the red star and triangle refer to the published H values for HgSe and CdSe [46, 47]. (Reproduced from [45]. CC BY 4.0.)

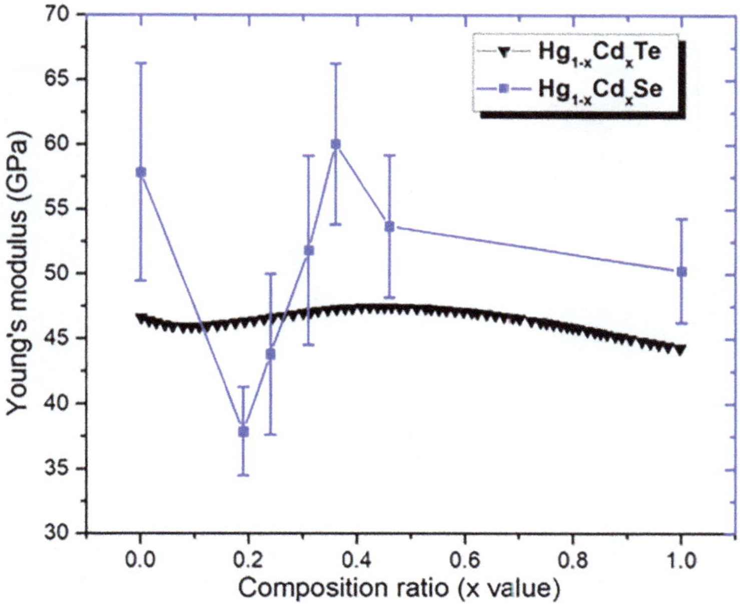

Figure 6.12. The measured E values (blue squares) as a function of x value for the $Hg_{1-x}Cd_xSe$ material, and the calculated E values (black triangles) based on the Bagot model for the $Hg_{1-x}Cd_xTe$ material. (Reproduced from [45]. CC BY 4.0.)

attributed to the formation of short-range segregated clusters [50], which hinder dislocation movement and reduce the likelihood of cross-slip events, thereby enhancing mechanical stability [51]. This hardening effect reflects the influence of local bonding and microstructural evolution as Cd substitutes Hg within the lattice. The experimental data generally conform to Vegard's law, as indicated in figure 6.11, suggesting a near-linear dependence of hardness on Cd composition across the range studied.

With nanoindentation experiments, the average elastic modulus values (E_{avg}) were calculated for each composition, yielding 57 ± 8, 38 ± 3, 44 ± 6, 52 ± 7, 60 ± 6, 54 ± 5, and 53 ± 4 GPa for $x = 0$, 0.19, 0.24, 0.31, 0.36, 0.46, and 1, respectively [44, 45]. These results are plotted as a function of composition in figure 6.12. It is observed that the variation of elastic modulus with increasing x-value in the $Hg_{1-x}Cd_xSe$ system resembles that of $Hg_{1-x}Cd_xTe$ alloys which can be described with the Bagot model [52]. Although this model was originally developed for the $Hg_{1-x}Cd_xTe$ system, the similar trends observed in the $Hg_{1-x}Cd_xSe$ experimental data suggest that the underlying mechanical behavior may be governed by analogous atomic-scale bonding mechanisms.

6.4 Demonstration of mid-wave infrared HgCdSe detectors [36]

To understand their IR detector applications, photoconductive detectors were fabricated based on high-quality HgCdSe materials grown on GaSb substrates.

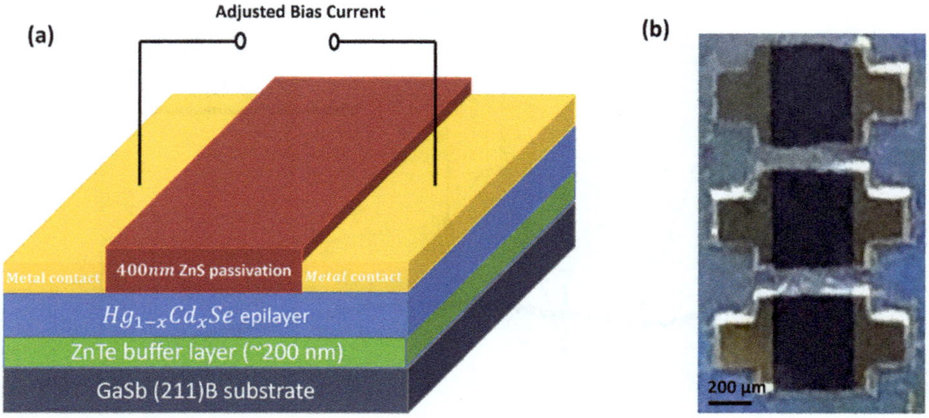

Figure 6.13. (a) Schematic device structure of HgCdSe photoconductor; (b) optical microscope image of HgCdSe photoconductors fabricated in this work. (Reproduced from [36]. CC BY 4.0.)

Figure 6.13 shows the device architecture of the HgCdSe photoconductors fabricated. Standard photolithographic techniques were employed to define square-shaped detector mesas with lateral dimensions of 600 μm × 600 μm. Following lithographic patterning, a selective wet etching process using a 0.5% Br_2/HBr solution was applied to etch the HgCdSe material and define the mesa structures. Subsequently, metal contacts were formed by thermal evaporation of Cr/Au bilayers with thicknesses of 20 and 200 nm, respectively. To provide surface passivation and reduce surface-related recombination, a 400-nm-thick ZnS layer was thermally evaporated onto the devices, completing the device fabrication process.

Figure 6.14 shows the measured spectral responsivity of the fabricated MWIR $Hg_{0.73}Cd_{0.27}Se$ photoconductive detector at 77 K. Two distinct photoresponse peaks are observed in the spectrum: the first with a cut-off near 1650 nm, and the second with a cut-off wavelength at around 4200 nm. The shorter-wavelength response, terminating near 1650 nm, is attributed to the Te-doped GaSb substrate, which is plausibly caused by the failure of electrically isolation of ZnTe buffer layer. As a result, photons with energies above the GaSb bandgap (~0.748 eV at 77 K, corresponding to ~1650 nm) can be absorbed by the substrate, contributing to the total photoconductive signal. The longer-wavelength peak, extending to a cut-off at ~4200 nm, corresponds to the intrinsic spectral response of the $Hg_{0.73}Cd_{0.27}Se$ active layer. This cut-off is consistent with the material's bandgap and aligns well with the spectral cut-off obtained from the FTIR transmission measurements, confirming the optical activity of the HgCdSe layer in the MWIR range.

The detector exhibits a peak responsivity of approximately 40 V W^{-1} at 77 K. While this value is considerably lower than that of conventional low-doped MWIR HgCdTe photoconductive detectors (typically on the order of 10^4 V W^{-1}), the reduced performance is attributed to the relatively high native n-type doping concentration in the MBE-grown HgCdSe epilayer ($>10^{17}$ cm^{-3}). This high background doping concentration increases background conductivity and reduces photoconductive gain, thereby limiting detector responsivity.

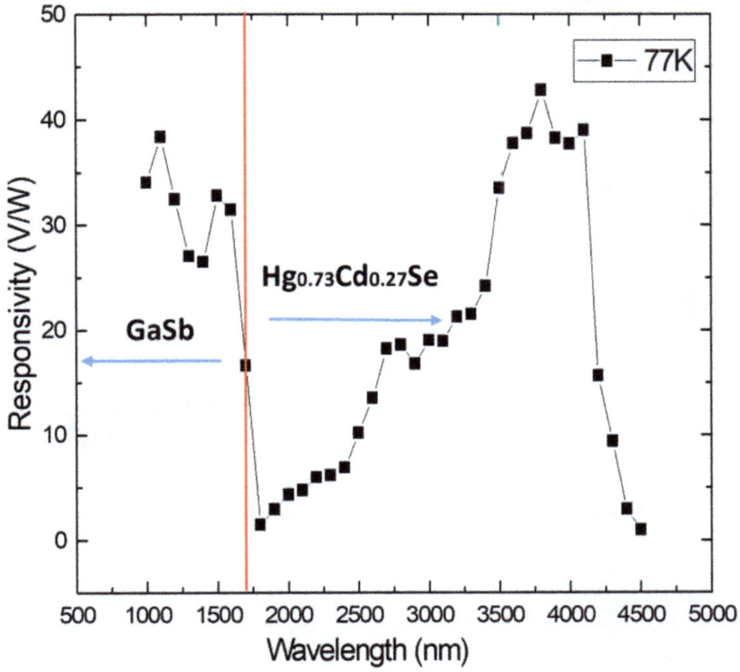

Figure 6.14. 77 K responsivity spectrum of MWIR $Hg_{0.73}Cd_{0.27}Se$/ZnTe/GaSb (211)B photoconductor device. (Reproduced from [36]. CC BY 4.0.)

Figure 6.15 shows the noise spectrum of the MWIR $Hg_{0.73}Cd_{0.27}Se$/ZnTe/GaSb (211) B photoconductive detector at 77 K. A characteristic $1/f$ noise behavior is observed at low frequencies with a transition (knee point) occurring near 2500 Hz. Beyond this frequency the spectrum flattens, indicating dominance by generation–recombination (g–r) noise. The spectral features, including the $1/f$ noise behavior and the g–r plateau, are consistent with those reported for MWIR HgCdTe photoconductive detectors [53]. Using the measured responsivity and noise spectrum, the nominalized specific detectivity (D_λ^*) of the $Hg_{0.73}Cd_{0.27}Se$ photoconductor was calculated according to the D^* expression discussed in chapter 1:

$$D_\lambda^* = \frac{R_\lambda}{V_n}(lw\Delta f)^{0.5}, \tag{6.10}$$

where R_λ is the peak responsivity, V_n is the total noise voltage, l and w are the length and width of the detector active area, and Δf is the noise bandwidth (1 Hz in this case). Substituting the measured parameters into equation (6.10), the calculated detectivity of the MWIR $Hg_{0.73}Cd_{0.27}Se$ photoconductor is approximately 1.2×10^9 cm $Hz^{1/2}$ W^{-1} at 77 K. This performance is comparable to that of MWIR $InAs_{0.91}Sb_{0.09}$ photoconductive detectors ($\sim 6.1 \times 10^9$ cm $Hz^{1/2}$ W^{-1}) [54], but remains two orders of magnitude lower than that of optimized, low-doped MWIR HgCdTe detectors ($\sim 2.0 \times 10^{11}$ cm $Hz^{1/2}$ W^{-1}) [55].

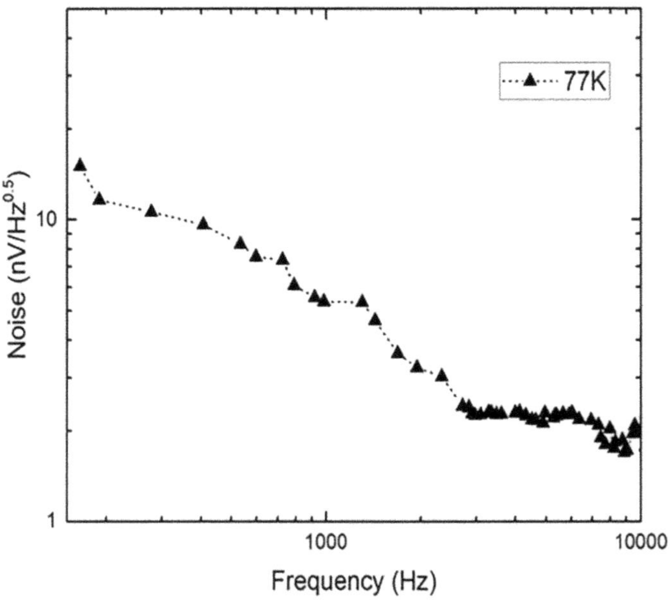

Figure 6.15. 77 K noise spectrum of MWIR $Hg_{0.73}Cd_{0.27}Se$/ZnTe/GaSb (211)B photoconductor device. (Reproduced from [36]. CC BY 4.0.)

In all, compared with its HgCdTe counterparts, the $Hg_{0.73}Cd_{0.27}Se$ photoconductor presents a relatively lower performance which can be primarily attributed to its higher background electron concentration and thus degraded electron mobility and minority carrier lifetime. As demonstrated in previous work on HgCdTe [56], reducing the background doping level not only enhances mobility by suppressing carrier–carrier scattering and screening effects, but also increases minority carrier lifetime by mitigating Auger and trap-assisted recombination. Therefore, the high background carrier concentration in HgCdSe materials has constituted a critical barrier for achieving high-performance HgCdSe IR detectors. As demonstrated in previous work [1], postgrowth thermal annealing in a Se environment can significantly reduce the Se vacancies, and thus effectively reduce the background doping concentration. One order of magnitude reduction in background electron concentration has been demonstrated for HgCdSe after postgrowth annealing in a Se environment [1]. In addition, higher purity Se source material can also effectively reduce the level of impurities in the HgCdSe epilayer, and thus the background doping concentration [1]. It should be noted that only 5N5 purity Se source material was used for growing HgCdSe at UWA. Therefore, it is expected that the background electron concentration in the HgCdSe grown at UWA can be reduced to be within the low-10^{15} cm^{-3} range or even lower by using higher purity (6N or above) Se source material and by implementing a post-growth thermal annealing process in a Se environment.

6.5 Summary

In a summary, HgCdSe IR materials grown on GaSb substrates provide an alternative material system for developing lower-cost, larger-array format and high performance IR detectors. As discussed in this chapter, HgCdSe IR materials present similar IR properties to their HgCdTe counterparts such as wide tuneable cut-off wavelength (bandgap), high electron mobility, and long minority carrier lifetime, making them suitable for developing high-performance IR detectors. Over the past decade, significant progress has been made in the area of MBE growth of HgCdSe on GaSb substrates. By optimizing the thermal desorption process of GaSb substrates, ZnTe buffer layer thickness, and HgCdSe growth conditions, high-quality HgCdSe materials have been achieved such as LWIR $Hg_{1-x}Cd_xSe$ ($x = 0.18$) materials with an XRD FWHM of 65 arcsec, an electron mobility of 1.3×10^5 cm^2 V^{-1} s^{-1} at 80 K (with a background electron concentration of 1.6×10^{16} cm^{-3}), and a minority carrier lifetime of 2.2 μs at 80 K. These electrical properties of LWIR $Hg_{1-x}Cd_xSe$ ($x = 0.18$) materials are comparable to those of the counterpart LWIR HgCdTe materials. Despite these, the current major challenge for the IR detector applications of HgCdSe materials is the high background electron concentration, which can change their optical bandgap and degrade their electrical properties and thus, the resultant detector performance. In the future, various techniques should be studied to reduce the background electron concentration to be within the low-10^{15} cm^{-3} range or even lower, including postgrowth thermal annealing in a Se environment, higher purity Se source material (6N or above) and others. Once properly controlled background electron concentration is achieved, advanced detector architectures should be studied such as p–n diode, nBn architecture, and others, to fully leverage the excellent physical properties of HgCdSe materials to make high-performance IR detectors with features of lower cost and larger array format size.

References

[1] Lei W 2018 A review on the development of GaSb alternative substrates for the epitaxial growth of HgCdTe *J. Nanosci. Nanotechnol.* **18** 7349–54

[2] Doyle K, Swartz C H, Dinan J H, Myers T H, Brill G, Chen Y P, VanMil B L and Wijewarnasuriya P 2013 Mercury cadmium selenide for infrared detection *J. Vac. Sci. Technol.* B **31** 03C124

[3] Zhang Z K, Pan W W, Liu J L and Lei W 2019 A review on MBE-grown HgCdSe infrared materials on GaSb (211)B substrates *Chin. Phys.* B **28** 018103

[4] Lei W, Ren Y L, Madni I, Umana-Membreno G A and Faraone L 2018 MBE growth of high quality HgCdSe on GaSb substrates *Infrared Phys. Technol.* **92** 197–202

[5] Brill G, Chen Y and Wijewarnasuriya P 2011 Study of HgCdSe material grown by molecular beam epitaxy *J. Electron. Mater.* **40** 1679–84

[6] Li Y H, Walsh A, Chen S Y, Yin W J, Yang J H, Li J B, Da Silva J L F, Gong X G and Wei S H 2009 Revised *ab initio* natural band offsets of all group IV, II-VI, and III-V semiconductors *Appl. Phys. Lett.* **94** 212109

[7] Maimon S and Wicks G W 2006 nBn detector, an infrared detector with reduced dark current and higher operating temperature *Appl. Phys. Lett.* **89** 151109

[8] Lansari Y, Cook J and Schetzina J 1993 Growth of HgSe and $Hg_{1-x}Cd_xSe$ thin films by molecular beam epitaxy *J. Electron. Mater.* **22** 809–13

[9] Chen Y P, Brill G, Benson D, Wijewarnasuriya P and Dhar N 2011 MBE growth of ZnTe and HgCdSe on Si: a new IR material *Proc. SPIE* **8155** 320–5

[10] Brill G, Chen Y and Wijewarnasuriya P 2011 Material characteristics of HgCdSe grown on GaSb and ZnTe/Si substrates by MBE *Proc. SPIE* **8155** 326–34

[11] Zhao W F, Brill G, Chen Y and Smith D J 2012 Microstructural characterization of HgCdSe grown by molecular beam epitaxy on ZnTe/Si(112) and GaSb(112) substrates *J. Electron. Mater.* **41** 2852–6

[12] Chai J, Lee K-K, Doyle K, Dinan J and Myers T 2012 Growth of lattice-matched ZnTeSe alloys on (100) and (211) B GaSb *J. Electron. Mater.* **41** 2738–44

[13] Madni I 2017 Characterization of MBE-grown HgCdTe and related II-VI materials for next generation infrared detectors *PhD Thesis* (The University of Western Australia)

[14] Lei W, Gu R J, Antoszewski J, Dell J, Neusser G, Sieger M, Mizaikoff B and Faraone L 2015 MBE growth of mid-wave infrared HgCdTe layers on GaSb alternative substrates *J. Electron. Mater.* **44** 3180–7

[15] Brill G N, Chen Y, Wijewarnasuriya P S and Dhar N K 2012 Hg based II–VI compounds on non-standard substrates *Phys. Status Solidi (A)* **209** 1423–7

[16] Chai J, Noriega O, Dedigama A, Kim J, Savage A, Doyle K, Smith C, Chau N, Pena J and Dinan J 2013 Determination of critical thickness for epitaxial ZnTe layers grown by molecular beam epitaxy on (211) B and (100) GaSb substrates *J. Electron. Mater.* **42** 3090–6

[17] García J N, ĎElía R, Heredia E, Geraci A, Tolley A, Di Stefano M, Cabanillas E, Martínez A and Trigubo A B 2015 Crystalline quality of CdSe single crystalline commercial wafer *Procedia Mater. Sci.* **9** 444–9

[18] Brill G and Chen Y 2011 Development of 6.1 a materials for IR applications *Project Report (ARL-TR-5855)* (US: Army Research Lab)

[19] Wenisch J, Schirmacher W, Wollrab R, Eich D, Hanna S, Breiter R, Lutz H and Figgemeier H 2015 Evaluation of HgCdTe on GaAs grown by molecular beam epitaxy for high-operating-temperature infrared detector applications *J. Electron. Mater.* **44** 3002–6

[20] Becker C R and Krishnamurthy S 2011 Chapter 12 - Band structure and related properties of HgCdTe *Mercury Cadmium Telluride: Growth, Properties and Applications* ed P Capper and J W Garland (West Susses: Wiley) pp 275–95

[21] Krishnamurthy S, Chen A-B, Sher A and Van Schilfgaarde M 1995 Temperature dependence of band gaps in HgCdTe and other semiconductors *J. Electron. Mater.* **24** 1121–5

[22] Guenzer C S and Bienenstock A 1973 Temperature dependence of the HgTe band gap *Phys. Rev.* B **8** 4655

[23] Whitsett C R, Broerman J G and Summers C J 1981 Crystal growth and properties of $Hg_{1-x}Cd_xSe$ alloys *Semiconductors and Semimetals: Defects, (HgCd)Se, (HgCd)Te* ed R K Willardson and A C Beer (New York: Academic) pp 53–118

[24] Chandrasekhar Rao T, Antoszewski J, Faraone L, Rodriguez J, Plis E and Krishna S 2008 Characterization of carriers in GaSb/InAs superlattice grown on conductive GaSb substrate *Appl. Phys. Lett.* **92** 012121

[25] Carmody M, Edwall D, Ellsworth J, Arias J, Groenert M, Jacobs R, Almeida L, Dinan J, Chen Y and Brill G 2007 Role of dislocation scattering on the electron mobility of n-type long wave length infrared HgCdTe on silicon *J. Electron. Mater.* **36** 1098–105

[26] Swartz C, Tompkins R, Giles N, Myers T, Edwall D, Ellsworth J, Piquette E, Arias J, Berding M and Krishnamurthy S 2004 Fundamental materials studies of undoped, In-doped, and As-doped $Hg_{1-x}Cd_xTe$ *J. Electron. Mater.* **33** 728–36

[27] Cui H, Zeng J, Tang N and Tang Z 2013 Analysis of the mechanisms of electron recombination in HgCdTe infrared photodiode *Opt. Quantum Electron.* **45** 629–34

[28] Doyle K, Swartz C H, Pattison J, Chen Y P and Myers T H 2013 Electron transport and minority carrier lifetime in HgCdSe *2013 II–VI Workshop (Chicago, Illinois, October 1–3)*

[29] Gu R, Lei W, Antoszewski J and Faraone L 2016 Investigation of substrate effects on interface strain and defect generation in MBE-grown HgCdTe *J. Electron. Mater.* **45** 4596–602

[30] Dasilva F W O, Raisin C, Silga M, Nouaoura M and Lassabatere L 1989 Chemical preparation of Gasb (001) substrates prior to MBE *Semicond. Sci. Technol.* **4** 565–9

[31] Kodama M, Hasegawa J and Kimata M 1985 Influence of substrate preparation on the morphology of GaSb films grown by molecular beam epitaxy *J. Electrochem. Soc.* **132** 659

[32] Silva F D, Raisin C, Silga M, Nouaoura M and Lassabatere L 1989 Chemical preparation of GaSb (001) substrates prior to MBE *Semicond. Sci. Technol.* **4** 565

[33] Zazo L G, Montojo M, Castano J and Piqueras J 1989 Chemical cleaning of GaSb (1, 0, 0) surfaces *J. Electrochem. Soc.* **136** 1480

[34] Schwartz G P, Gualtieri G, Griffiths J, Thurmond C and Schwartz B 1980 Oxide-substrate and oxide-oxide chemical reactions in thermally annealed anodic films on GaSb, GaAs, and GaP *J. Electrochem. Soc.* **127** 2488

[35] Wang C, Shiau D and Lin A 2004 Preparation of GaSb substrates for GaSb and GaInAsSb growth by organometallic vapor phase epitaxy *J. Cryst. Growth* **261** 385–92

[36] Zhang Z K, Pan W W, Umana-Membreno G A, Ma S, Faraone L and Lei W 2025 MBE growth of high quality HgCdSe for infrared detector applications *Materials* **18** 3676

[37] Capper P, Harris J E, O'Keefe E S, Jones C L and Gale I 1995 Macro-and microsegregation of Zn in bridgman-grown CdZnTe *Adv. Mater. Opt. Electron.* **5** 101–8

[38] Chai J, Noriega O C, Dinan J H and Myers T H 2012 Critical thickness of ZnTe on GaSb (211) B *J. Electron. Mater.* **41** 3001–6

[39] Pan W, Zhang Z, Lei W and Faraone L 2019 Optical properties of MBE-grown $Hg_{1-x}Cd_xSe$ *J. Electron. Mater.* **48** 6063–8

[40] Summers C and Broerman J 1980 Optical absorption in $Hg_{1-x}Cd_xSe$ alloys *Phys. Rev.* B **21** 559

[41] Madni I, Membreno G A U, Lei W and Faraone L 2018 Investigation of MBE-growth of mid-wave infrared $Hg_{1-x}Cd_xSe$ *J. Electron. Mater.* **47** 5691–8

[42] Chu J, Qian D and Tang D 1986 Burstein-Moss effect in $Hg_{1-x}Cd_xTe$ *Phys. Scr.* **T14** 37–41

[43] Chu J, Li B, Liu K and Tang D 1994 Empirical rule of intrinsic absorption spectroscopy in $Hg_{1-x}Cd_xTe$ *J. Appl. Phys.* **75** 1234–5

[44] Zhang Z, Pan W, Martyniuk M, Ma S, Faraone L and Lei W 2022 Nanoindentation of $Hg_{0.7}Cd_{0.3}Se$ prepared by molecular beam epitaxy *Infrared Phys. Technol.* **127** 104446

[45] Zhang Z, Pan W, Martyniuk M, Ma S, Faraone L and Lei W 2024 Determination of elasto-plastic properties of semiconducting $Hg_{1-x}Cd_xSe$ using nanoindentation *Infrared Phys. Technol.* **136** 105057

[46] McCumiskey E J, Chandrasekhar N and Taylor C R 2010 Nanomechanics of CdSe quantum dot–polymer nanocomposite films *Nanotechnology* **21** 225703

[47] Gorai S K 2012 Estimation of Bulk modulus and microhardness of tetrahedral semiconductors *J. Phys. Conf. Ser.* **365** 012013

[48] Shchennikov V V, Ovsyannikov S V and Frolova N Y 2003 High-pressure study of ternary mercury chalcogenides: phase transitions, mechanical and electrical properties *J. Phys.* D **36** 2021

[49] Andrews R, Walck S, Price M, Szofran F, Su C-H and Lehoczky S 1990 Microhardness variations in II-VI semiconducting compounds as a function of composition *J. Cryst. Growth* **99** 717–21

[50] Zax D, Vega S, Yellin N and Zamir D 1987 Study of structural ordering in $Hg_{1-x}Cd_xTe$ by 125Te NMR *Chem. Phys. Lett.* **138** 105–9

[51] Fissel A and Schenk M 1990 Microhardness of $Hg_{1-x}Cd_xTe$ and $Hg_{1-x}Zn_xTe$ *Cryst. Res. Technol.* **25** 89–95

[52] Bagot D, Granger R and Rolland S 1994 A comparison of force constants, mechanical and thermal properties in $Hg_{1-x}Cd_xTe$ and $Hg_{1-x}Zn_xTe$ mixed crystals *Phys. Status Solidi (B)* **183** 395–406

[53] Sewell R, Musca C A, Dell J M, Faraone L, Jozwikowski K and Rogalski A 2003 Minority carrier lifetime and noise in abrupt molecular-beam epitaxy-grown HgCdTe heterostructures *J. Electron. Mater.* **32** 639–45

[54] Tong J, Xie Y, Ni P, Xu Z, Qiu S, Tobing L Y and Zhang D-H 2016 $InAs_{0.91}Sb_{0.09}$ photoconductor for near and middle infrared photodetection *Phys. Scr.* **91** 115801

[55] Siliquini F, Musca C, Nener B and Faraone L 1995 Temperature dependence of $Hg_{0.68}Cd_{0.32}Te$ infrared photoconductor performance *IEEE Trans. Electron Devices* **42** 1441–8

[56] Yoo S D and Kwack K D 1998 Analysis of carrier concentration, lifetime, and electron mobility on p-type HgCdTe *J. Appl. Phys.* **83** 2586–92

Chapter 7

Outlook for next-generation HgCdTe infrared detectors with features of lower cost and larger array format

As the last chapter of this book, chapter 7 summarizes the whole book by reviewing the content, analysis, and discussion, as well as conclusions of all the previous chapters, 1–6. It also presents the outlook/perspective for developing next-generation infrared detectors with features of lower cost and larger array format size by discussing the current/future challenges and potential solutions of growing high-quality HgCdTe on alternative substrates and HgCdSe on GaSb substrate.

7.1 Summary of previous chapters

Chapters 1–6 have thoroughly reviewed and discussed the development of heteroepitaxial growth of HgCdTe and HgCdSe, especially lattice-mismatched growth of HgCdTe, for making high-performance IR detectors with features of lower cost and larger array format size. Chapter 1 presented a brief introduction on IR technology and HgCdTe IR detectors, as well as their future challenges, leading to the topic of this book—*Lattice-mismatched Epitaxy for Fabricating HgCdTe Infrared Materials and Detectors* to address the challenge of higher cost and smaller array format size of HgCdTe IR detectors. Chapter 2 presented a theoretical discussion on the growth mechanisms related to lattice-mismatched heteroepitaxy including conventional 3D on 3D-mismatched heteroepitaxy and 3D on 2D-mismatched heteroepitaxy. Various general approaches were also proposed from the theoretical point of view to reduce the defects in the epilayers grown. Chapter 3 reviewed and discussed the progress of heteroepitaxial growth of CdTe and HgCdTe on various lattice-mismatched alternative substrates including Si, Ge, GaAs, and GaSb substrates. Detailed experimental studies were discussed and various approaches were proposed to control and

doi:10.1088/978-0-7503-3443-3ch7 © IOP Publishing Ltd 2025. All rights, including for text and data mining (TDM), artificial intelligence (AI) training, and similar technologies, are reserved.

annihilate the dislocations generated, especially threading dislocations. Chapter 4 reviewed and discussed the progress of vdW epitaxy of CdTe and HgCdTe on various 2D substrates including graphene, mica, and other TMD monolayer substrates. Chapter 5 reviewed and discussed the progress of HgCdTe IR detectors grown on various alternative substrates including SWIR, MWIR, and LWIR detectors. Chapter 6 reviewed and discussed the heteroepitaxy of HgCdSe on GaSb substrates to replace HgCdTe for making high-performance IR detectors with features of lower cost and larger array format.

In all, significant progress has been made in the area of lattice-mismatched heteroepitaxy of HgCdTe on alternative substrates. Various alternative substrates have been studied such as Si, Ge, GaAs, and GaSb, and various dislocation reduction techniques have also been proposed and studied to control and annihilate the dislocations within. However, most of the CdTe and HgCdTe layers on lattice-mismatched substrates still present a high EDP level of mid-10^6 cm^{-2} to low-10^7 cm^{-2}, which can be tolerated by SWIR and MWIR HgCdTe detectors, but will significantly deteriorate LWIR detectors' performance due to their much narrower energy bandgap. It should be noted that with some dislocation reduction techniques the EPD numbers in HgCdTe grown on GaAs and CdTe grown on GaSb were already demonstrated to be below the critical EPD level (5×10^5 cm^{-2}) for making high-performance LWIR HgCdTe IR detectors. Therefore, it is promising to achieve high-quality HgCdTe IR materials on alternative substrates with an EDP below the critical level of 5×10^5 cm^{-2}. But significant effort is still needed to apply those dislocation reduction techniques to real industry manufacturing. Apart from HgCdTe on alternative substrates, the heteroepitaxy of HgCdSe on GaSb also presents a potential pathway to make high-performance IR detectors with features of lower cost and larger array format size. However, high background doping concentration constitutes a challenge to the ultimate applications of HgCdSe grown on GaSb. Further effort, especially industry participation, is needed to develop this new HgCdSe IR technology towards ultimate industry applications.

7.2 Outlook/perspective for next generation HgCdTe infrared detectors with features of lower cost and larger array format

Looking ahead, it is still promising to achieve high-quality HgCdTe IR materials on alternative substrates with an EDP below the critical level of 5×10^5 cm^{-2}. By comparing the discussions in chapters 3 and 4, MBE growth of CdTe and HgCdTe on alternative substrates (Si, Ge, GaAs and GaSb) is more straightforward than that grown on 2D substrates, and results in much higher material quality for the CdTe and HgCdTe layers grown. Despite the various benefits of vdW epitaxial growth, CdTe and HgCdTe layers grown on graphene and mica present relatively poor material quality. Therefore, in the short term, MBE growth of CdTe and HgCdTe on alternative substrates (Si, Ge, GaAs, and GaSb) is more promising than the vdW epitaxial growth. But more effort is required to further control and reduce the dislocation density in the CdTe and HgCdTe layers to a level comparable to that of HgCdTe grown on lattice-matched CdZnTe substrates. How can this be achieved? Apart from further optimizing those dislocation reduction techniques discussed before, there are two main challenges which are usually neglected, but will be the major challenges to be addressed:

(1) *Lattice mismatch between CdTe and HgCdTe:* Although the lattice mismatch between CdTe and HgCdTe is tiny, it can have a significant impact on the HgCdTe material quality as HgCdTe layers are usually thicker (>5 μm), and exceed the critical thickness. Therefore, although some alternative substrates are reported to have a very low EPD for the CdTe layers, the subsequent HgCdTe layers still present an EPD in the range of mid-10^6 cm^{-2} to low-10^7 cm^{-2}. So it is essential to address this tiny lattice mismatch. Some potential techniques to address this include: composition graded buffer between CdTe and HgCdTe, transitional buffer layer to better relax the mismatch between CdTe and HgCdTe, and others.

(2) *Industry compatible dislocation reduction process:* Although cyclic thermal annealing has led to an EPD in HgCdTe grown on GaAs of around 2.3×10^5 cm^{-2}; no related industry manufacturing process has been reported since. Such cyclic thermal annealing is a quite complex process which might be difficult to reproduce. Therefore, a simpler dislocation reduction process might be more suitable for industry manufacturing.

As discussed before, apart from HgCdTe on alternative substrates, the heteroepitaxy of HgCdSe on GaSb also presents a potential pathway towards making high-performance IR detectors with features of lower cost and larger array format size. However, currently there are only two institutions (US ARL and UWA) working on this new HgCdSe IR technology as reported in the open literature. Compared with HgCdTe technology, the lack of strong industry participation seriously slows down the overall development of this HgCdSe IR detector technology. The main reason might be that the IR industry hasn't seen the real potential of HgCdSe to compete and replace HgCdTe. In this case, significant further effort is needed to show the full potential of HgCdSe by demonstrating prototype detector devices with performance comparable to their HgCdTe counterparts. However, to demonstrate the full potential, there are several major challenges to be addressed:

(1) *High background doping concentration:* As discussed in chapter 6, high background electron concentration (a level of 10^{16}–10^{17} cm^{-3}) can change the optical bandgap and degrade the electrical properties and thus ultimate detector performance of HgCdSe. To make high-performance IR detectors, the background electron concentration should be controlled to be within the low-10^{15} cm^{-3} range or even lower. Some potential techniques to address this include postgrowth thermal annealing in a Se environment, higher purity Se source material (6 N or above), and others.

(2) *Low temperature detector processing:* As HgCdSe materials on GaSb are typically grown at low temperatures (<100 °C), this places stringent

requirements on the device processing temperature. The relevant detector processing recipes developed for HgCdTe must be modified and redeveloped to ensure they will not damage the HgCdSe device and deteriorate the detector performance.

(3) *Advanced detector architectures:* Apart from photoconductor and photodiode, barrier detector structure, for example, the HgCdSe–ZnTe–HgCdSe nBn structure, provides an ideal device architecture for achieving high-performance HgCdSe IR detectors (potentially higher operating temperature). However, the large growth temperature difference between the HgCdSe and ZnTe barrier layer presents a challenge to growing nBn device structure with a high material quality. Some new growth techniques should be studied to grow a ZnTe barrier layer with a material quality sufficient for making high-performance nBn detectors such as two-step growth (low temperature growth of thin ZnTe as capper, then standard high temperature growth of ZnTe to have high material quality), and others.

In either case—HgCdTe on alternative substrates or HgCdSe on GaSb substrate—significant effort is still needed to further develop these technologies for ultimate industry manufacturing. The success will lead to high-performance HgCdTe and/or HgCdSe IR detectors with features of lower cost and larger array format, which will significantly broaden their industry applications.

www.ingramcontent.com/pod-product-compliance
Ingram Content Group UK Ltd.
Pitfield, Milton Keynes, MK11 3LW, UK
UKHW051041120926
8556IPUK00030B/564

9 780750 334419